FOREST PRODUCT AND THEIR UTILIZATION

FOREST PRODUCT AND THEIR UTILIZATION

Praveen Taank

CYBER TECH PUBLICATIONS
4264/3, Ansari Road, Daryaganj, New Delhi-110002 (India)
Ph.: 011-23244078, 011-43559448 Fax: 011-23280028
E-Mail: cyberpublicationsdelhi@yahoo.com
Website: www.cybertechpublications.com

FOREST PRODUCT AND THEIR UTILIZATION

Praveen Taank

First Edition 2010

Asian Offset Printers

Published by :

G.S. Rawat for **Cyber Tech Publications**
4264/3, Ansari Road, Daryaganj, New Delhi-110002 (India)
Ph.: 011-23244078, 011-43559448 Fax: 011-23280028
E-Mail: cyberpublicationsdelhi@yahoo.com
Website: www.cybertechpublications.com

Preface

Forest closely links human activities with nature. Similar to agriculture, forests also reveal the wealth of a nation. Investment in forestry and realisation of returns is taken care of by the state, due to the long gestation period and several laws governing the exploitation of forest resources. However, efforts are taken by the government to widen the forest cover through several afforestation programmes and agroforestry projects. The forest development have to understand certain basic principles underlying forest management.

An attempt is made in this book to explain the principles of economics and management as applied to a development project in the forest sector. However, communities vary in their interest and capacity to manage their tree and forest resources. Why have some communities planted or managed their trees and forests well for generations while others have not? Why have some working management systems recently broken down but others have continued?

Often investigation shows that when forest management is successful there are sets of rules and expectations related to tree use and management that the community respects. Where incentives for the community to follow these rules have been weakened, management often breaks down. In order to stimulate increased community efforts to improve and manage trees and forests, forestry activities must recognize and support local rules and expectations and provide incentives for local investment in tree and forest resources. A better understanding of how to better support local institutions in forest management.

This book is prepared with the undergraduate students of forestry in mind and hence the sophisticated mathematical applications are limited. It is believed that this book will be useful to them.

—*Author*

CONTENTS

Preface

1. Forest Products 1
2. Role of Forestry and Trees in the National Economy 58
3. Forest Biodiversity Values and Functions 81
4. Forest Productivity 108
5. Utilization of Forest Products 152
6. Forest Based Industrial Development 185

Bibliography 238

Index 239

1

FOREST PRODUCTS

INTRODUCTION

Important NWFPs of India, their importance in rural and industrial economy: Significance of indigenous NWFP for human welfare, industrial and environmental management; non-conventional uses; recycling of pre and post harvested products.

Fiber yielding species: Species diversity, phytogeographical distribution; fiber substitutes, local and region uses.

Grasses, sedges and reeds: Natural and production areas, agro-climatic areas of distribution, grasses and reeds (Phragmites, Arundo, Arundinela) of commercial use; Grasses used in rural economy and uses; Munj; Sabai; Fodder yielding grasses. Silvipastoral- cum- NWFP significance.

Bamboo and canes: Diagnostic features of identification; Parts used as NWFP, species diversity, Endemic centres and priority sites of bamboo and rattan; Distribution patterns of sporadic and gregarious flowering; Tribal areas of traditional use of species; Grass root innovation on utilization; Criteria and indicators of sustainable management, National Application on Bamboo Mission.

Oil Seeds: Diagnostic features of oil yielding NWFPs; Forest and Agro forestry origin, Agro climatic, Bioclimatic distribution; Species diversity and salient features for identification; Priority species selection; Adulterants.

Gums, resins and oleoresins: Plants gums and resins; Bioclimatic and agroclimatic distribution of species; Edible and non edible yielding species; lesser known species outside forest; Traditional and commercial uses; Criteria for selection of species as additive.

Tannins and Dyes: Indigenous and exotic species; natural and production areas; Distribution patterns diagnostics; Edible and controlling dyes of plant and no plant origin. Lesser known, local and traditional uses; Substitutes for adulteration and commercial utilization.

Edible plants: Classification; distribution and salient features of lesser known and under utilized species of NWFP potential; Food from forest; Emergency and lean season food plants; Bioclimatic distribution; Diagnostic features of uses.

PRACTICAL

i. Survey, identification and characterization using traditional and integrating recent approaches.
ii. Design and layout of NWFP gardens.
iii. Species selection, criteria, indicators and diagnostics for utilization
iv. NWFP inventorization and monitoring of an potential area
v. Characterization and categorization of threatened species providing, NWF Products for utilization.
vi. NWFP impact assessment with emphasis on loss of resources, substitution by adulterants, trade, demand and supply and cost benefit analysis.
vii. Development of systematic database on the NWFP resource for conservation and sustainable.

Nursery & Plantation Technology

1. Types of nursery, layout, important prerequisites, different types of nursery beds, sunken and raised beds, their layout, size of the beds, fencing, providing shades in the nursery.
2. Irrigation system: Sprinkles system, Economics of seedling production.
3. Seed collection, processing and storage, recalcitrant and orthodox seeds, germination percent, seed treatment, seed testing.
4. Modern nursery techniques, Containerized seedling production system.
5. Macro propagation and micro propagation.
6. Green House, Mist Chamber, Shade House, Construction and usefulness in seedling production, different designs and their costs estimates.
7. Tissue culture seedling production system, Hardening of seedling.
8. Fertilizers and its application, growth hormones and its applications.
9. Disease and pest control in nursery.

Practical

i. Preparation of nursery beds, seed treatment, different methods of seed sowing, seed testing, seed processing and storage.

ii. Preparation of Plan/ Project document on establishment of nursery, construction of green house, mist chamber and composting unit, cost estimation.

iii. Macro propagation, Air layering, Grafting.

iv. Hardening of seedling

v. Preparation of medium for root trainers, sowing of seed, transplanting of seedlings, watering.

vi. Application of fertilizers, preparation of micronutrients solutions.

vii. Soil testing- pH, NPK and assessment of organic carbon.

viii. Production of large sized seedlings

CHEMISTRY OF NON WOOD FOREST PRODUCTS

Oil Seeds

Different types, edible and non-edible, physio-chemical properties, isolation and characterization of fatty oils and their uses.

Gums, Resins and Oleoresins

Different type of gums, isolation of gums, their physical properties and uses; types of resins, different methods of resin tapping, oleoresins: general chemistry and application.

ESSENTIAL OIL

General methods of isolation, characterization and uses.

Isolation and Characterization of Fatty Oils

Practical

i. Estimation of fatty oil and essential oil.

ii. Isolation of natural dyes and dying trials.

iii. Preparation of the plant extracts.

iv. General methods for the isolation of pure compounds from plants

v. Isolation of seed gums from plants and determination of their viscosity.

MEDICINAL & AROMATIC PLANTS

1. Herbal medicines- History and current research.
2. Medicinal plant resources of India.
3. Collections of MAPs from forests, harvesting, processing, grading and storage.
4. Cultivation techniques of selected MAPs economics of cultivation. Atees, sarpgandha, geranium, safed musali, giloy, isabgol, jaropha.
5. Status of National and International trade of medicinal and aromatic plants.
6. Sustainable management of MAP resources, in-situ and ex-situ conservation.

National Medicinal Plant Board and its Activities

Practical

i. General methods of isolation and identification of compounds from medicinal plants.
ii. Cultivation techniques of important species. Study Tour

Project Assignment: Finalization of topic, project formulation, statistical methods and designs for laying out experiments and collection of Field data.

Value Addition of Important NWFPs

1. General principles, methods and technique employed in the value addition of Medicinal & Aromatic plants (MAPs), fatty oils, gum, resins, oleoresin, tannins.
2. Aroma therapy, isolation of fatty acids and essential oils, isolation of natural dyes and dying methods, general methods for isolation of pure compounds, isolation of seed, gums and determination of their

rheological properties.

3. Value addition of bamboo, vinegar, pickle, soap making and handicrafts.
4. Methods of charcoal making from wood and bamboo, various types of kiln used for manufacture of charcoal, charcoal dust briquettes.
5. Miscellaneous products: Bidi leaves, leaf fodder, animal products.

Practical

i. Isolation of fatty acids and essential oils.
ii. Isolation of natural dyes and dying methods.
iii. General methods of isolation of pure compounds.
iv. Isolation of seed, gums and determination of their rheological properties.
v. Preparation of aroma therapy products: decoction, tincture, syrup, concentrates and absolutes.
vi. Katha and cutch form Acacia catechu

PRODUCT WORK

Controversy about forest destruction has been the Bank's Achilles Heel for nearly two decades. The flagship disasters in Brazil and Indonesia in the 1980s focused on the role of the Bank as forest destroyer. The subsequent attempts by the Bank to do good forestry through the TFAP (Tropical Forest Action Plan) proved a further public relations disaster for the Bank. The TFAP was an ambitious plan launched by the World Bank and other international actors in the mid-1980s to mobilize US $ 8 billion over a period of five years. While the official goal was to curb deforestation, the TFAP's planned investments were very much geared towards industrial logging activities. It was the uniting of ecological justice NGOs in opposition to the TFAP that created the WRM.

The 1991 Policy was an important step forward for the Bank. It committed the Bank to a new approach to forests including commitments to:

- assess the likely impacts on forests of macro-economic and non-forest sector loans: SAPs, agricultural projects, dams, mines etc.
- take into account the interests of communities in establishing plantations
- involve forest-dwelling people in establishing protected areas
- avoid funding logging in primary moist tropical forests
- condition other lending for forest operations on the inclusion of measures to safeguard the interests of forest-dwellers
- experiment with new projects with forest-dwellers including in promoting non-wood products.

Forest Policy Implementation Review, partly as a result of insistence of NGOs and a supportive US Executive Director. This was a rather unsuccessful stock-taking exercise, although the scope given for NGO participation in the process was a step forward in Bank procedures.

A half-hearted attempt by some Bank staff to have the proscription on funding of logging in primary moist tropical forests was lifted by sustained NGO pressure. The WRM was especially critical of the failure of the review to look at:

- the cross-sectoral impacts of Bank lending on forests
- the lack of staff incentives to observe the policy, especially with respect to adequate participation
- the social impacts of Bank lending related to forests, especially in terms of impacts on women, indigenous peoples and land tenure, or measures to address these impacts.

The Bank admitted that these were real deficiencies in its review but that it was obliged to report to the World Bank's Board of Executive Directors (EDs) that year and so had no time to fill the gaps. The WRM heavily lobbied the EDs at the Bank not to accept the Review as it stood. In the end they did accept it on condition that a follow up review was carried out in 1997. After many delays this revised review is now underway.

FOREST DWELLERS

- 'Experimental programs to test alternative approaches to the participation of local people and to the promotion of nonwood products of natural forest to benefit such people will also be financed' (Sector:65).
- Financing of forestry projects will depend on the borrower government adopting a 'comprehensive and environmentally sound forestry conservation and development plan that clearly defines the roles and rights of government, the private sector and local people (including forest dwellers).

DEGRADED OF THE FORESTS

An IRS-1D multispectral image is used to measure the extent to which central forests are degraded. The study area covers most of the eastern half of Uttaranchal state in the Indian Himalayas. Accuracy assessment was performed on a sample taken from an Ikonos 1-meter resolution image. We estimate that 61% (48-73%) of the forested area has less than 40% crown cover, with numbers in brackets denoting the 90% confidence interval. The calculation of confidence intervals is a novel feature of the study that is important in view of the fact that satellite image interpretation is never free of error. The results show that the Forest Survey of India's widely cited figures considerably understate the extent of degradation.

Over the past few decades there has been a great deal of concern about ecological degradation and deforestation in the Himalayas. The main aim of this paper is to provide estimates of the area of forest degraded in two Himalayan districts with 90% confidence intervals by using satellite imagery.

Degradation is measured in terms of crown cover. An IRS-1D multispectral LISS-III satellite image from May 1998 that covers an area of 20,000 square km in the Indian central Himalayas northwest of Nepal, is analysed. We present a vegetation map and estimates of the percentage of the area under forest or scrub, and the percentage of the forested area that is degraded. These figures are presented along with 90% confidence intervals, derived from crown cover counts done on a sample from an Ikonos 1-m resolution panchromatic image from April 2000.

The Forest Survey of India's biennial State of Forest Reports1, based on interpretations of IRS imagery, have been the main source of district-wise data on the state of forests in India.

However, their estimates are presented without confidence limits or error matrices, and no information about the interpretation method is given. This study finds that a greater proportion of the forest area in two districts of Uttaranchal is degraded than the Forest Survey of India estimates would suggest.

Forest degradation has resulted from a number of human uses. Overgrazing, lopping for firewood and fodder, resin tapping in the presence of fire, and commercial felling have all contributed to degradation. Degradation reduces crown cover and can proceed to the point when only scrub is left. Since, except for small areas of juniper above the treeline, scrub does not occur naturally in the area, we treat scrub as degraded forest.

The degradation of the forests has been a major blow to rural residents of the area who have suffered from the lack of firewood (the main source of energy for cooking and heating), fodder and grazing, and the adverse effects on springs which are the main water sources for villages on the slopes. Degradation may also affect areas in the plains since it may increase the runoff during the monsoon, leading to increased flooding during the monsoon months and reduced water supplies at other times of the year. Finally, of course, forests play a role in carbon fixation.

Methods

The aim was to obtain estimates of the area under forest or scrub and the percentage of the forest area degraded, together with confidence intervals for these estimates. A two-stage classification process was applied to the IRS image. In the first stage, training sites acquired during two years of fieldwork in 1997-99 were used to separate pine from broadleaf forests and forests from other classes.

This was done using a supervised maximum likelihood classification into 11 classes which were subsequently concatenated. In the second stage, quantitative estimates of canopy cover from the Ikonos imagery were used to separate the two forest types into dense and degraded classes. The Ikonos data were used to obtain sample error matrices for the classifications. These were used to compute producers' accuracies, that is the probabilities of classifying pixels from a given true class in the various classes. Point estimates of the true class proportions were obtained by inverting the producer's accuracy matrix and multiplying it by the vector of class proportions from the classified map. Confidence bounds for these point estimates were obtained by bootstrapping the sample error matrices and associated estimated true class proportions.

POINT ESTIMATES OF DEGRADATION

The producer error matrix is defined by P = [pjk], where pjk is the probability that a pixel that is truly in class k gets classified in class j. From the vegetation map, the shares of land s = (s1,..., s5) under five classes (dense and degraded broad-leaved and pine forests, and grassland and agriculture) in the IRS image were obtained. The other classes either represented tiny fractions of the area like bare land, or (like snow) had negligible confusion with other classes, and so were excluded from the error matrix. Now,

s = Ps*,

where s* is the unknown vector of true shares of land in each class. Therefore, the vector of true shares of land in each class is given by

s* = P-1s.

Let P ^ = [p^ jk], where p^ jk is the proportion of sample pixels really in class k that were classified as class j. Replacing P by P ^ yields the following estimate, due to ref. (6), of the true shares of land in each class:

s^ * = P ^-1s.

s^

* is a consistent (roughly speaking, unbiased in large samples) estimate of s* if P^ is a consistent estimate of P. P is a consistent (in fact, unbiased) estimate of P, provided the misclassification probabilities in the Ikonos image area from which the sample was drawn are the same as the misclassification probabilities for the area as a whole, and for the subregions of interest.

This is our assumption. Since the vegetation types are the same and the Ikonos image covers a transect of the area, this assumption is thought to be reasonable, although, of course, it is unlikely to be wholly accurate. The more narrowly each class is defined, the more likely it is that the assumption will hold. On the other hand, expanding the number of classes raises the cost of getting a sample of a given size for each class, and also raises the probability of infeasible estimates (class shares that are less than zero or greater than one). With these considerations in mind we chose to use five classes.

The consistency of P does not require a simple random sample. It is sufficient that each subsample from a given true class be random. This is important since we would not have been able to obtain enough points in the rarer classes with a simple random sample. Note that it is certainly not true that the actual shares of land in each class are the same for the sampled area and the areas of interest.

This is why it is not possible to use the alternative adjustment procedures for estimating s* proposed by Tenenbein and Card. Those methods rely on consistent estimates of user accuracies ujk, that is, the probability that a pixel mapped as class j was really from class k. To get consistent estimates of user accuracies, the sample must be taken from an area with representative true class shares. This can be difficult for the following reasons.

Reference data usually come from either field observations or high-resolution imagery. Field samples are expensive. A randomly chosen point may prove to lie on the boundary of different true classes when visited in the field. This is especially likely to be the case in tropical or subtropical montane areas with high variability in landuse.

Using such a point is likely to result in the confusion of registration error with classification error. Dropping such points raises the cost of getting a sample of given size. Highresolution images (as in our study) are often available only for portions of the area of interest, and, therefore, are not representative of the shares of land use classes in the area of interest. Finally, if there is more than one area of interest, as in our study, methods utilizing the user error matrix would require a sample from each of these areas, thus multiplying the cost of the study.

The drawback of the method used here, which uses the producer error matrix, is that it requires a matrix inversion. This means that s^

* could have elements that are less than zero or greater than one. This problem is unlikely to occur if the diagonal elements of P (the probabilities that classes are correctly classified) are sufficiently close to l, and if the class proportions are not too unequal.

Matrices with diagonal elements too close to zero are illconditioned, meaning that small changes in the elements of such a matrix can lead to large changes in its inverse. Since it is P ^-1 that is used in the calculation of the true class proportions, if it is ill-conditioned, then the estimated true class proportions will have a large variance. A simulation study found that if the producer accuracies (that is, the diagonal elements) were at least 0.7, then this problem did not occur. This is why obtaining a classification accuracy of 70% in each class is desirable but higher accuracies are not necessary, provided the assessment sample size in each class is large enough.

Some of the diagonal elements in the error matrix are less than 70%, resulting in considerable variation in the inverse matrix. However, as we shall see below, we can still conclude that degradation has been overestimated by the Forest Survey of India. Thus, despite the low classification accuracies in some classes, we can still reach useful conclusions because the confidence intervals inform us about the limits of the likely errors.

Confidence bounds for the degradation estimates

We now come to the main contribution of this paper: computing confidence intervals for our estimated proportions of the areas in the different classes. s^

* is a useful estimate only if we know how much error it is likely to contain. The 90% confidence intervals on the elements of s^

* were computed as follows. One thousand random 'bootstrap' samples with replacement were drawn from the sample of 553 points used for accuracy assessment above, each being of the same size as the original. (See ref. 12 for a discussion of bootstrap methods in the context of estimating error rates in classification problems.) Each bootstrap sample is thus a random draw from the original sample and results in a different distribution of band ratios in each true class. Then s^

* was calculated for each bootstrap sample by re-calculating P ^-1 and multiplying it by the original map proportions vector s. The 5th and 95th percentiles of the frequency distribution of s^

* thus generated give us the 90% confidence intervals on s^

*. The most straightforward way to narrow the confidence intervals and increase the precision of the estimates is to use a larger sample for those classes for which the sample size is small. This will reduce the sampling variability between bootstrap samples and thus narrow the confidence intervals on the error matrix, and therefore, on the proportions of the area in different classes. An increased sample size may also improve classification by allowing better estimates of the parameters although this effect is likely to be small except when starting from a very small sample size. Of course, increasing the sample size is costly in terms of increased

field visits, or acquisition, geo-registration, and visual interpretation of high-resolution images.

Results

Referring to Table 3, we see that 72% (68-77%) (throughout the paper the ranges refer to the 90% confidence interval) of the old Almora district, subsequently split into Almora and Bageshwar, was under forest or scrub in 1998.

The Forest Survey of India (FSI) estimated the area under forest or scrub to be much lower: only 48%. As mentioned in the introduction, the FSI estimates are presented without any information on the interpretation methods used and without confidence bounds.

In fact, the FSI estimated the area under scrub in Almora district to be only about 1% of the total area, an unbelievably low number. This probably accounts for much of the wide discrepancy between their estimates and ours. Of the forest (including scrub) area, we find that 78% (68-86%) was degraded, i.e. had a crown cover of less than 40%. The FSI estimates the forest degraded to be just 19%. Again, a good part of the discrepancy is probably due to their misclassification of scrub and other degraded forests as non-forest.

A caveat that needs to be made at once is that our IRS image does not cover the whole of both districts (Figure 1). Nevertheless, this cannot account for such huge discrepancies. The part of Almora district not in our image is very small. The part of Pithoragarh left out is larger, but most of this is under permanent snow.

The FSI's underestimation of the forest and scrub area and related underestimation of the proportion of forested area that is degraded is repeated in the Pithoragarh district (subsequently split into Pithoragarh and Champawat). We estimate the forest or scrub area in the Pithoragarh district to be 71% (67-75%) while the FSI figure is only 36%. Again, their figure for the per cent of the total area of the district under scrub is rather low: just 4%. We find that 58% (45-70%) of the forested area has a crown cover below 40% while the FSI estimate is 31%.

While there is one other large-scale study that uses visual interpretation of 1972 Landsat MSS data and aerial photographs for ground truths13, it is unfortunately not directly comparable with ours, since they used different crown cover cutoffs and made no attempt to distinguish between scrub and grasslands. There are also a number of studies of much smaller areas in the region 14-17. They are all based on aerial photographs from the early 1970s and they generally found smaller proportions of forest with high crown cover than we do. Tiwari and Singh17 also included a time-change analysis using recent IRS imagery and found a decline in the extent of forests with high crown cover.

CLASSIFICATION AND ACCURACY ASSESSMENT

Ground features, compass, altimeter, and a GPS was used to locate the training site on a map. Each site was marked as a polygon leaving a 30-m buffer of the same vegetation around the polygon to eliminate errors in the signatures arising from edge effects and registration error. After eliminating sites that were in shadow or had positional errors, 169 sites, each consisting of one of eight classes, were selected for training. These were:

(1) dense broad-leaved forest,

(2) open broad-leaved forest,

(3) broad-leaved scrub

(4) dense pine forest,

(5) open pine forest,

(6) grassland,

(7) agriculture, and

(8) bare land.

The broad-leaved forest with a basically closed canopy was classified as dense. Conifers, mainly pine, were classified as dense if it was not possible to see the ground when the forest was looked at from the outside, and if the trees were close enough to leave no gaps between them. Signatures for the classes

(9) water bodies,

(10) snow, and

(11) cloud, were taken directly from the image.

Since the registration error from attempting to georeference the whole IRS image was high, those parts of the satellite image containing the training sites were georeferenced with root mean square registration errors ranging from 0.45 to 1.03 pixels.

The image was classified using the method of maximum likelihood assuming that the four bands were jointly normally distributed with different parameters for each class, and with each class given a prior probability of being equally likely. The entire procedure was implemented with Erdas Imagine 8.4 software. Classes (1)-(3) of those mentioned above, were merged to form a class called broadleaved forest (BL), (4) and (5) were merged to form Pine forest, (6) and (7) were merged into a category called Grasses, and the others were merged to form a category called Other.

After this study was underway, Ikonos 1-m resolution imagery became available. This offered an opportunity to improve the measurement of forest degradation since aerial photographs of this region are not in the public domain and were not available to us. An image from 24 April 2000, covering a transect 30 km in length from north to south and 2.5 km wide was procured from among the few images available for the area (Figure 1). This was intended to span the altitudinal zones corresponding to different vegetation types in the region. The Ikonos image was registered to the IRS image with a root mean square error of 0.7 IRS pixels.

After a field visit to prepare a key for visual interpretation, a random sample of 1024 points was taken from the transect. 75 75 m windows around these points were visually interpreted to correspond with the classes described above. The purpose of the 75 75 m windows (roughly equivalent to 3 3 IRS pixels) was to ensure that the true class of an IRS pixel at a point was correctly determined.

A smaller window might have led to true classes being incorrectly determined due to registration error. An additional sample of 256 points was randomly taken from areas rich in vegetation types under-represented in the original sample. These areas were determined visually from the Ikonos image, so that the additional points were selected on the basis of the reference or true class, and not the map class.

Six hundred and eight points for which the entire 75 75 m window fell in a single class, were retained for analysis. If the 75 75 m window contained more than one class, then we could not confidently assign the true class of the IRS pixel containing the random point, since registration error might lead to a shift in the location of the IRS pixel on the Ikonos image.

A producer error matrix for this sample, P = [pjk], where pjk is the percentage of points in class k that were classified by the maximum likelihood procedure as class j. Here the true class k refers to the 75 75 m window about the point, while the map class j refers to the class of the IRS pixel containing the point. This error matrix presents only three classes, broad-leaved forest, pine forest, and grassland and agriculture. The Other class is not presented since its confusion with forest classes is negligible. Table 1 shows that separation of the three vegetation types is good, with over 80% producer accuracy being achieved for all three.

It was found that the first stage classification done above, while differentiating quite well between different forest types and grasslands and agriculture, performed poorly in distinguishing between dense and open classes in the same type of forest. We resorted to the second stage procedure described below to separate dense and open classes. A 10-m grid was laid on the Ikonos image and a 7 7 segment of this grid centred on each random point in the forest classes was used to measure crown cover.

The proportion of the 49 squares in each segment of the grid that were covered by tree crowns was the measure of crown cover. Of the 608 points, 553 points for which reliable crown cover counts were possible were retained.

Crown cover was used as the indicator of degradation. Any cutoff level for classifying an area as dense or open, is, of course, arbitrary. Two cutoffs were used: 40 and 80%. Eighty per cent was thought to be appropriate since an undisturbed broad-leaved forest in this region would have a closed canopy2. The 40% cutoff was used for comparison with the data generated by the Forest Survey of India, which uses a 40% cutoff.

The data were topographically normalized by using the ratios of band 5 to bands 2, 3 and 4. Classification into dense and degraded classes for both cutoffs for both broad-leaved and pine types was done using the maximum likelihood classifier applied to the three band ratios and aspect. Aspect is defined as a continuous variable taking all values from 0 for south-facing pixels to 1 for north-facing pixels.

The aspect variable was created in ArcMap 8 software using freely downloadable elevation data with a 90 m resolution from the SRTM (Shuttle Radar Topography Mission). (We are grateful to a referee for suggesting the use of aspect data for improving the classification accuracy which resulted in narrower confidence intervals.) The prior probabilities were chosen to roughly maximize the sum of classification accuracies of dense and degraded types while not allowing any of them to fall below 50% in the assessment sample.

Notice that there is still considerable misclassification of open or degraded categories as dense and vice versa, even after the use of the second-stage procedure.

The fact that the different forest types are classified with a fairly high degree of accuracy, but separation of degraded from dense forest is subject to considerable error is only to be expected.

There is not much difference between forests with crown cover slightly less than 40% and slightly more than 40%. There is no natural sharp distinguishing feature and so considerable misclassification is unavoidable in the latter case. For the purpose of estimating areas in the various classes, however, this need not matter as long as the producer error matrices are estimated with sufficient accuracy.

FOREST CONSERVATION

It is stating the obvious to note that the native forest estate is increasingly the centre of considerable public controversy. The debate over forests is a microcosm of the broader debate surrounding environmental issues because it continually confronts society with complex choices.

Each choice offers a variety of paths to the future, and in turn, each path provides net 'value' to society. Net 'Value' is comprised of a mix of quantities and types of 'values' which individually and collectively meet the needs of society. Differing 'paths to the future' confront society with the need to estimate the net value likely to be derived from each route. In economic analysis, and (it is assumed) society at large, the path which offers the most value to present and future citizens is the preferred choice.

Many complex social choices are decided in the realm of politics with some choices arbitrated on a more technical, arms length basis by designated officials. Choices in environmental policy historically combine technical, 'objective', elements and qualitative, 'subjective', elements. As a result, the choices are subject to a range of pressures that may not beset other types of decisions. In this context, it follows that estimating the value, or total 'economic welfare', of different paths within environmental policy may be more difficult than the simpler financial trade offs common to economic decision making. 'Optimising the benefits to the community' has been identified as a key objective in forest policy. In order to achieve this goal, it is necessary to identify the contribution that different policy options can make to the welfare of the community. In other words, we must determine which path to the future offers the greatest 'net value', and which combination of values acts to create that path.

This discussion examines the relationship between financial values derived from timber harvesting, the financial values derived from alternative uses of the forest (eg. water, tourism etc), the financial values derived from plantation timber and the non financial values of the forest (eg. environmental amenity) through the reviewing of a series of reports on native forests.

THE CONTRIBUTION OF ECONOMIC THEORY

The Ecologically Sustainable Development Working Group on Forest Use specified three main requirements for sustainable forest use:

- maintaining the ecological processes with in forests (the formation of soil, energy flows, and the carbon, nutrient and water cycles);

- maintaining the biological diversity of forests;
- and optimising the benefits to the community from all uses of forests within ecological constraints. The National Forest Policy Statement adopts these principles as the basis for ecologically sustainable development. (National Forest Policy Statement (NFPS), 1992:47) (Emphasis added.)

Economic theory sets a standard by which the outcomes are measured, and informs us as to what are 'good' and 'bad' ways of making choices. Economic theory also provides an important underlying philosophy which specifies that changes should be to the net benefit of society at large.

The Role of Economic Theory

Economic theory is often seen as a method, or science, of analysing impacts on the financial e c o n o m y. When properly and rigorously deployed, however, it is elevated to the 'art ' of choice. Thorough economic analysis concerns itself with how to make choices that provide the greatest net benefit to the community, where 'benefit' implies consideration of all social values. It is for this reason that economic theory, incorporating such concepts as externalities and public goods, has come to play a significant role in decision making processes. The following sections briefly outline the relationship between the concepts of externalities, economic value and economic decision making.

Externalities and the Welfare of Society

Economics is centred on the allocation of scarce resources to competing ends, in order to achieve the most value for human society. In this instance, economic analysis seeks to identify the flows of value that can be derived from 'uses' and 'non uses' of the forest.

This process of analysis seeks the combination of uses and non uses that most meets the value preferences of society, both now and in the future. The process is intended to be value neutral in the sense that it is based on the given or expected values of society-whatever those values may be.

Traditionally, economic theory uses the 'market' as its benchmark of efficient resource allocation; that is, an allocation of resources that creates the greatest benefit for society. This 'pure' model, however, admits that its results hold true only under very tight and specific conditions. If any of these conditions are breached, then there can be no presumption that the free market outcome is the best one.

Externalities arise when certain conditions of a free market do not hold. They arise when the free market fails to include, or take account of goods and services provided incidentally by the completion of another transaction. An externality can be compared to an auction where the main transaction is completed between two parties, but where additional goods and services, not provided for in the original transaction, are subsequently delivered.

These can be delivered either to the main parties or to third parties who were not involved in the original deal, and who, indeed, may even be unaware of the original transaction. For example, if a tree is cut down in the forest for timber, the loss of that tree may adversely affect the view of a home owner, who subsequently experiences a loss in his/her property value.

Externalities and public goods are both expressions of market failure. 'Public goods' refers to a value from whose benefit others cannot be excluded. Thus, deriving pleasure from knowing of the existence of a forest is an externality, because others cannot be excluded from deriving the same pleasure. At the same time, the phenomenon is also a public good because it is non rival in consumption; that is, many can consume the item without reducing the consumption of others.

In order to avoid confusion, all such values are called 'non financial values' since they cannot, by definition, be traded in a market place, and hence be expressed in terms of financial value. Where environmental issues are concerned, the market cannot be relied upon to produce an optimal allocation of resources-ie. the greatest overall benefit to the community-because there is no venue, no 'global auction room', in which the community (in particular future generations) can express its preferences over the items it values.

The 'items of value' in this discussion are the 'values' or 'benefits' that flow from the 'use' of 'objects'. Normally, economic analysis talks of objects such as goods and services when the real interest is in the values that flow from these objects. It makes little sense to talk, for instance, of a 'forest' per se. Are we referring to the wood alone, the biological species that reside there, the pleasure that bushwalkers derive, or its function in absorbing greenhouse gases? The term clearly has no operational meaning unless we refer explicitly to its various use values and benefits. In order to avoid any ambiguity, the remainder of this paper will generally use the term 'values' rather than 'goods' or 'products' (as 'values' is a more inclusive term).

An understanding of externalities is central to quality economic decision making, particularly in the context of the environment-simply because in the real world, there are many items of value (fresh air, knowing that an old growth forest exists, etc) that cannot be bought and sold in the market. The exclusion of these items from the 'economic equation' means that the overall equation is miscalculated. Miscalculations of the total economic equation generally lead to a misallocation of economic resources, and to an overall loss to society. The practical significance of identifying externalities and public commodities relates to their impact on markets. It is impossible for a market alone to systematically produce the socially optimal outcome when externalities and public commodities are present, as they are in the case of the natural environment 1.

Pricing Theory and the Financial Values of the Forest

In economic theory, the price of an item can represent one of three circumstances for the owner.

- The price could reflect the value of the item to the owner.
- The price could value the item at a level greater than the owner's subjective valuation. In this case the owner is likely to sell the item.
- Alternatively, the price of the item may be less than the value to the owner in which case the owner will refuse to sell.

The price of wood must also reflect the different type of values generated by the wood and the portion of forest which is altered by its extraction.

Hence, if the 'public citizen' 2 offers a logging coupe for sale, by definition the value anticipated from the sale must at least exceed the value expected from any alternative use. And additionally, the price needs to equal:

- the costs of the production of wood;
- plus a return on the capital invested, with sufficient reward to cover the risk; and
- total compensation for all non financial values (NFV's) lost in the process of an extraction of the timber.

This pricing method ensures that timber with values exceeding the market price will not be sold. However, the significance of NFV's is that they make up a considerable proportion of the value of timber. When NFV's are destroyed, the loss of value is not felt by one person alone, but is likely to be felt by every Australian citizen, every international citizen and everyone yet to be born. When logging occurs, each of these people experiences a change in the value they derive from the forest, and hence each should be compensated. The price of timber needs to be able to compensate all losers if policy is to be socially optimal.

The pricing of timber is crucial to determining the pattern of logging in the forest. In order to implement a solution, the present approach of forestry commissions uses regulation to separate logging areas from non logging areas. This approach could be assuming one of two things. Firstly, it could be assuming that royalties are sufficient compensation for the loss of NFV's 3. In this case, it remains to be demonstrated by those who set royalty levels that they have evaluated the worth of the lost NFV's, and thus set royalty levels accordingly. In the absence of information about how NFV's are determined, it could be assumed that non financial values in the logged areas are considered insignificant to the extent that the Australian community will not require compensation for the losses in NFV's which result. In theory, logging will then occur, in the areas set aside, if the cost of production and capital makes it a financially rational exercise.

Pricing policy is, thus, conducted on a 'compartmentalised basis'. In other words the forest is surveyed for its values, and those areas where NFV's are deemed 'small enough' to avoid the need for compensation, are made available for logging. The remainder is put aside for conservation. Thus, if we use a 'compartmentalised' approach, timber is valued in a two stage process. The forest is allocated to different uses depending on the scale of NFV's in different areas. Forest that is to be subject to logging is then priced on a commercial basis. Pricing takes place on the assumption that the coupes have already been surveyed, and the NFV's have been found to be minor; ie very close to zero.

The worth of the royalties, of course, must also cover the costs of wood production, the return on capital required, and the financial opportunity cost of foregone uses. The Victorian Auditor General's report (1993) indicated the likelihood that the forestry operations in Victoria were making low profits (if any at all). Given that royalties must cover operating, capital and opportunity costs, this leaves very little room to compensate for lost NFV's. In such a circumstance, it is feasible that NFV's are being valued at a worth very close to zero. No evidence was identified to justify such a valuation.

THE THEORY OF TOTAL ECONOMIC VALUE

In any public decision over the net benefits of forests, externalities and public commodities, along with project and alternative financial values, need to be taken into account, as part of the value provided by a 'resource'. In analysing different resource options, a total listing and weighting of values, both private and 'externalised', must be provided for each option.

When an economist is comparing alternative social arrangements, the proper procedure is to compare the total social product yielded by these different arrangements... In devising and choosing between social arrangements we should have regard to the total effect. This, above all, is the change in approach which I am advocating.

Coase refers to 'total social product' which we have renamed 'total economic value' in the tradition established by Pearce (1989). In order to compare the total economic value of differing policy options, it is necessary to aggregate for each option all the values (financial, aesthetic, spiritual, etc.) that would be produced for the community if a particular policy option were to be undertaken.

Traditional project evaluation normally compares one option against another option, and chooses that which has the most value. In the presence of market imperfections such as externalities, the total economic value (the project's total costs and benefits to society) can be severely miscalculated in at least two ways. First, proper economic evaluation needs to consider the financial impact that one project may have on other parties, not simply whether it is profitable when considered in isolation. Suppose, for example, that the government is requested to give its approval for the establishment of a factory on a river bank. Viewed in isolation, the proposal is attractive: it is expected to provide jobs for 100 local unemployed young people and will contribute an extra $5 million to tax revenues. Viewed from a wider perspective, however, the project's net financial value would be seen to be negative if the poisonous effluent it emits cause the bankruptcy of downstream fish farms and tourist resorts which together generate 200 jobs and $10 million in tax revenues. These 'down stream' financial values are often overlooked by traditional project evaluation methodologies.

The second weakness of the traditional evaluation approach is that it gives little consideration to non financial values. To continue with the illustrative example above, the factory's pollution may poison wildlife and household pets, ruin a picnic spot favoured by thousands of local residents, etc.

These types of costs are real and relevant, notwithstanding the fact that they are of a 'non financial' nature. Resolving the first problem is easier, since it may be possible to obtain the relevant financial data if appropriate economic modelling is conducted. The second problem is more difficult because non financial values are not traded in markets. In other words, there is no 'objective dollar price' which can provide a yardstick with which to compare non financial and financial values. Environmental economists have been working on providing techniques to evaluate non financial values.

Economic Decision Making

As a society, and as individuals, we are constantly evaluating and rejecting options. A social cost benefit analysis is, in effect, the act of making choices. Social cost benefit analysis becomes a conceptual framework that expresses the less formal processes we use to make choices. While the National Forest Policy Statement leaves open the practical question of assessment, it identifies the need for evaluation of policy options. In Diagram 1 we have listed all the values, in the tradition of Pearce (1989) and Young (1992) that appear to be provided by the native forest. Table 1 describes each type of value, and where relevant, their relationship to each other. Total economic value (TEV) analysis would require that each policy option would receive a 'score' in each cell of an evaluation matrix (see Table 5 for an example). Scores are aggregated by a means yet to be determined which then enables decisions over differing options to be made.

Economic decision making seeks to choose between the differing options on the basis of determining that which provides the greatest net social benefit. The means of deciding which option has the greatest net social benefit where 'scores' cannot be aggregated into a single numeraire is, of course, highly contentious.

There are three channels through which resources can be allocated. These are the markets, institutional processes and executive decision making. Markets, however, as decision making institutions, are largely unavailable in the circumstances covered in this report because we are dealing specifically with instances of market failure. However, modified markets (ie. markets subject to government intervention) are included as a means of policy implementation. The choice between government modified markets and 'pure' market measures should be justified on the grounds of net social benefits. The irony is that if market measures are to be chosen as the decision making process, then it is a decision that will have to be made by government anyway, and hence does not relieve the public decision making process of the need to evaluate the optimality of value flows from the forests under differing policy options.

In other words, choosing to 'do nothing' is as much a decision as choosing to 'do something'. The lesson is that if optimal social benefits are sought, then market failure automatically triggers government decision making processes-even if that results in a decision to 'do nothing'.

The creation of a formalised process for considering environmental values and other externalities requires a clear understanding of the integrated structure of externalities, economic value and economic decision making.

All values-financial, ethical, ecological, etc-must be accounted for, if different policy options are to be compared on a equal basis.

The different values that comprise each policy option, need to be integrated in some manner which enables socially optimal choices to be made.

APPRAISING THE VALUES OF THE FOREST: THE THEORY AS PRACTISED

This section reviews thirteen recent reports on forests for their views on the appraisal of forest values. Each report was reviewed in order to gain an understanding of the value assessment problems in the forest sector. The reports were examined on the basis of four criteria:

- Identification of externalities.
- Identification of valuation methodologies.
- Identification of pricing/financial valuation practices.
- Identification of likely assessment frame works.

Problems in the Identification of Externalities

A representative example, of the values identified, is listed below. These values were derived from the Resource Assessment Commission (RAC) study. This study, amongst those examined, was the most comprehensive and thorough in its treatment of the issues. The weaknesses in its study were repeated, and magnified, in other studies.

The Resource Assessment Commission Inquiry into Forest and Timber

Conservation; wilderness; recreation; Financial values such as wood, tourism and water supply, oils, seeds, tannins, medicinal plants, sandalwood, bark products, fishing, hunting, wild foods, infrastructure corridors; Information via education, research; intrinsic values like aesthetic (different from visual), heritage, spiritual, biocentric, wilderness, lifestyle; Utilitarian values; 'pristinity'; landscape; scenic quality, public sensitivity, industry value by implication; existence values, option values, vicarious use values, bequest values, ecological sustainability, ethical values.

The main weaknesses in assessing the manner in which values had been surveyed were that:

- Values were not identified or listed in a systematic or comprehensive manner. The above extract from the RAC report was amongst the most comprehensive, but even so, lacked a systematic listing of values.
- Different value categories often overlapped. For example, ecological sustainability was repeatedly confused with 'environmental amenity'.
- There was also repeated confusion over the values we impute to an object, and the values we derive from that object as a result of our preferences. For instance, a forest site may have great scientific value, but not be of great aesthetic value. Conversely, a forest site

may have limited scientific value, but great aesthetic value. In the latter case, a purely scientific view point would not reveal the aesthetic importance of the site. Similarly, there are other values-for example existence value-held subjectively by the public with respect to native forests. A study which evaluated objective characteristics only would not pick up the subjectively felt values. We have identified this as the 'object subject' tension.

Confusion over the type and source of 'values' will undermine the efforts to assess all values. It is therefore necessary to have an agreed list of all values, comprehensively and systematically ordered, in order to achieve the goal of optimising benefits to the community. It is of note that 'Hoehn and Randall have shown that independent estimation of the component parts [of Total economic value], followed by their summation, is not formally identical to TEV estimated by a sequential procedure in which, say, existence value is estimated first, then 'site' experience value, then activity related values at the site.

This is because the consumer surplus for a bundle of goods is not the same as the sum of the consumer surpluses on individual goods.

Identifying Suggested Valuation Techniques

Each report suggested, implicitly or explicitly, a range of valuation techniques for providing an 'objective' assessment of the values identified. Overall, the lack of comprehensive identification of values, common to all thirteen reports, subsequently inhibited the matching of valuation techniques to appropriate categories of value.

This complicated the process of value assessment since it provided limited guidance about how various values should be measured.

On the positive side however, the Resource Assessment Commission (RAC, 1992) report did provide an assessment of each evaluation technique, which marks a very useful first step towards providing a listing of techniques relevant to the assessment of each value.

The Organisation for Economic Co-operation and Development references (OECD, 1986; 1992) also made a worthwhile contribution towards evaluating techniques.

In general, the reports provided only limited advice on the means by which different types of values should be evaluated. Moreover, with the exception of the RAC and the OECD reports, the techniques put forward were generally insufficient to measure the full range of values derived from the forest.

For example, the reports would discuss techniques for measuring the scientific values of the forest, with out discussing other measures that evaluate the 'worth' of the non-scientific values.

A second issue worthy of further discussion in each report was the role of uncertainty in complicating the assessment of values.

Whilst some discussion of uncertainty was entered into by four reports (RAC, 1992; ESD, 1992; AHC, 1992 and OECD, 1986), the issue was not raised sufficiently to provide guidance on how valuation techniques were to account for uncertainty. What is required is a more explicit discussion on the means by which 'uncertainty' should be handled in valuation processes.

The following table lists the valuation techniques that were identified. While there will be some debate about whether these are all 'techniques' or 'management strategies', it is clear that the implementation of these processes leads to the de facto valuation of forests.

For the purposes of this report, we have described them all as valuation techniques. Some of the techniques identified below were implicit in the various reports rather than explicitly described.

THE ECONOMIC DIMENSION OF TRANSITION

One way in which an economist familiar with cost-benefit analysis (CBA) may look at environmental impact assessment (EIA) is that EIA is a CBA requiring a large amount of information rigorously collect-ed and analyzed by a group of scientists of various disciplines. One can appreciate that the development of EIA has served to advance the systematic descrip-tion and quantification of environmental effects in a way which can only serve to improve the quality of CBA undertaken.

One can also appreciate that EIA constitutes a process, formalized through environ-mental legislation, regulations and procedures, which in turn reinforces the community participation re-quired for informed decision-makers seeking the common good.

While CBA within EIA is a key gen-erator of information to the decision-makers, recog-nition should also be given that the EIA process is both a planning tool and an execution monitoring tool.

Thus, the role of CBA is not as an intermediate step, but rather as an integrated parallel process.

Hence, the CBA should follow up, quantify and eval-uate the dynamic iterations created by the exhaustive search of alternatives, by the efficient inclusion of environmental protection and mitigation measures and by the explicit valuation of preference and choice ar-ticulated by community, society and their representa-tives acting in the best interest of present and future generations.

VALUE OF CONSIDERING ENVIRONMENTAL EFFECTS EARLY ON IN THE PROJECT CYCLE

The main purpose of the economic analysis of a project is to ascertain whether the project can be ex-pected to create more net benefits than any other, mutually exclusive option, including the option of not doing it. Consideration of alternative options, therefore, is a key feature in proper project analysis. Often, important choices about alternative project options are made early on in the project cycle.

These options may differ considerably in their general eco-nomic contribution, and they may also differ greatly concerning their environmental impact. Therefore, including environmental effects in the early econom-ic analyses, however approximately, should improve the quality of future decisionmaking.

The CBA process-defining objectives, searching for alternatives, costing out the resources involved, speci-fying the effects of each option involved and compar-ing all the costs and benefits-normally requires considerable efforts. In the case of Cuba, we are faced with two joint families of economic valuation prob-lems: the "environmental" ones and those problems associated with the incipient transition from a cen-trally planned economy to a truly participative mar-ket based economic system.

In principle, economic analyses are to take into ac-count all costs and benefits of a project. With regard to environmental impacts, however, there have been two basic problems even in developed countries.

First, environmental impacts are often difficult to measure in physical terms.

Second, even when im-pacts can be measured in physical terms, valuation in monetary terms can be difficult. In spite of such dif-ficulties, a greater effort needs to be made everywhere to 'internalizes' as many environmental costs and benefits as possible by measuring them in money terms and integrating these values in the economic appraisal.

The measurement in money terms are made even more difficult in countries undergoing market reforms.

The environmental valuation problems can be summarized as follows:

VALUE OF ENVIRONMENTAL ASSETS

While man-made and human capital may be valued with relative ease by observing existing market sys-tems, where available; the existence value of clean wa-ter and air, tropical forests, wetlands, coral reefs and other environmental assets and their functions is much more difficult since not even market prices can reflect their full contribution to other economic ac-tivity and to human welfare.

In particular, the mar-ket price of water does not reflect the various services it provides nor do market values can accurately re-flect what happens when irreversible loss or damage of natural resources occurs as environmental degrada-tion exceeds a critical threshold level.

Complexity of environmental valuation also arises due to the multi-ple functions of being a source of raw materials and energy, being a sink for assimilating man-made wastes and providing other services such as recre-ational/tourism services, storage of genetic diversity and scientific and educational benefits.

The following classification is useful:

- Direct use values are derived from the economic uses made of the natural system's resources and services. Examples of these are outputs such as timber, game and recreation from forests or fish and scuba tourism from coral reefs.

- Indirect use values are the indirect support and protection provided to economic activities and property by the resource system's natural func-tions or environmental services. Examples of these are watershed protection and soil erosion prevention provided by forests, and beach sand and mooring facilities protection provided by coral reefs.
- Non-use values lie in the special attributes of the natural system as a whole, its cultural and heri-tage uniqueness; it includes both existence and option values. Existence values reflect public goods which can be enjoyed by more than one consumer without decreasing the amounts en-joyed by others such as clean air, beaches and forests. The existence value is the utility that consumers derive from just knowing the public good exists. A way of measuring that utility would be to measure the willingness to pay or the contingent value assigned if the public good were to disappear. Option values reflect what current generations wish to bequeath for future generations to inherit. They imply both an ethi-cal commitment to sustainabilty for the children of our children and in a shorter time-frame the maintenance of options to solve current prob-lems. Examples of option values have been devel-oped for forests relating to biodiversity and the search for cures of cancer and AIDS-"if forest were to disappear then the options to find such cures will vanish" or "as long as the forest is pro-tected there is the option to find the cure".

Direct Effects Valued on Conventional Markets Some methods are directly based on market prices or productivity. This is possible where a change in envi-ronmental quality affects actual production or pro-ductive capability.

Change-in-Productivity: Development projects can affect production and productivity positively or neg-atively.

The incremental output can be valued by us-ing standard economic prices where available. Loss-of-Earnings. Environmental impacts can signifi-cantly affect human health. In theory, the value of health impacts should be determined by the willing-ness to pay of individuals to maintain their health. In practice, one uses earnings lost upon early death, dis-ease or job absence. This approach is used in highway and industrial safety, and in air pollution studies.

The "implicit value of human life" approach is reject-ed by many as dehumanizing since human life can be said to have infinite value. However, society, govern-ment regulations, insurance companies and judicial courts implicitly and explicitly place finite values on human life and health. This is a necessity reflecting limited resources to be allocated for health expendi-tures. The relatively high level of health expenditures in Cuba would indicate a high implicit value of hu-man life and health. However, one can also discern a political motivation and its attendant benefits behind it.

Preventive Expenditures: Individuals and governments invest in prevention measures to avoid or reduce un-wanted environmental effects. Environmental dam-ages, are often difficult to assess, but historical infor-mation on preventive measures and their costs may be interpreted as a minimum value for the expected benefits that the preventive measures seek. If it is found, for example, that there is a historical pattern of under investment in Cuba for natural disaster planning and prevention; then, it can be inferred that the benefits expected from such preventive measures have been very low. This conclusion would in turn imply that the loss of human life would be assigned a relatively low value. While this possible conclusion may appear to conflict with the high political priority for health expenditures, it can be observed that the level of damages or loss of lives caused by natural di-sasters cannot be easily attributed to government in-action and thus the political cost may be easily ex-plained away.

Potential Expenditure

Valued on Conventional Markets

Replacement Cost: Simply, the costs that would have to be incurred in order to replace a damaged asset. The estimate is not a measure of benefit of avoiding damage since the damage costs may be higher or low-er than the replacement cost.

Valuation Using Implicit (or Surrogate) Markets

Sometimes one must use market information indi-rectly. Approaches to be considered are the travel cost method, the property value approach, the wage dif-ferential approach, and uses of marketed goods as surrogates for non-marketed goods. Each technique has its particular advantages and disadvantages, as well as requirements for data and resources. One must determine which techniques might be applica-ble to a particular situation.

Travel Cost: This approach measures the travel cost (and time involved) which reflects the willingness of consumers or users can serve to measure the benefits produced by recreation sites (parks, lakes, forests, wilderness). It can also be used to value "travel time" in projects dealing with fuelwood and water collec-tion.

Property Value: This valuation method is based on the general land value approach and can determine the implicit prices of certain land areas. The property value approach can help analyze willingness to pay for properties with different pollution levels and infer the implicit cost of pollution. The method compares prices of houses in affected areas with equal size and similar neighborhood characteristics elsewhere in the same metropolitan area.

Valuation Using Constructed Markets

Contingent Valuation: When society's preferences as revealed in market prices are not available, the con-tingent valuation method tries to obtain information on individual preferences by posing direct questions about willingness to pay. It basically asks people what they are willing to pay for a benefit, and/or what they are willing to accept by way of alternate compensa-tion to tolerate an environmental cost. This process may be achieved through a direct questionnaire/ sur-vey. Willingness to pay is difficult to measure and depends on the income level of the sample subjects, and involves problems of designing, implementing and interpreting questionnaires. While its applicability may be limited, there is now considerable experience in evaluating the quality of supply of potable water and electricity services.

Artificial Market: Such markets can be constructed for experimental purposes, to determine consumer willingness to pay for a good or service. For example, a home water purification kit might be marketed at various price levels, or access to a game reserve may be offered on the basis of different admission fees, thereby facilitating the estimation of values placed by individuals on water purity or on recreation facilities.

THE DISCOUNT RATE

Discounting is the process by which costs and bene-fits occurring in different time periods may be com-pared. The discount rate to be used has been a gener-al problem in cost-benefit analysis, but it is particularly important with regard to environmental issues, since some of the associated costs and benefits are very long-term or irreversible in nature. In stan-dard analysis, past costs and benefits are treated as "sunk" and are ignored in decisions about the present and future. Future costs and benefits are discounted to their equivalent present value and then compared. In theory, in a perfect market, the interest rate re-flects both the subjective rate of time preference (of private individuals) and the rate of productivity of capital.

Higher discount rates may discriminate against fu-ture generations. This is because projects with social costs occurring in the long term and net social bene-fits occurring in the near term, will be favored by higher discount rates. It is often argued that discount rates should be low-ered to reflect long-term environmental concerns and issues of intergenerational equity. However, this would have the drawback that not only would eco-logically sound activities pass the cost-benefit test more frequently, but also a larger number of projects would generally pass the test and the resulting in-crease in investment would lead to additional envi-ronmental stress.

Many environmentalists believe that a zero discount rate should be employed to protect future genera-tions. However, employing a zero discount rate is in-equitable, since it would imply a policy of total cur-rent sacrifice, which runs counter to the proposed aim of eliminating discrimination between time periods-especially when the present contains wide-spread poverty.

In the case of projects leading to irreversible damage (e.g., destruction of natural habitats, etc.), the bene-fits of preservation may be incorporated into stan-dard cost-benefit methodology using the Krutilla-Fisher approach. Benefits of preservation will grow over time as the supply of scarce environmental re-sources decreases, demand (fueled by population growth) increases, and possibly, existence value in-creases. The Krutilla-Fisher approach incorporates these increasing benefits of preservation by including preservation benefits foregone within project costs. The benefits are shown to increase through time by the use of a rate of annual growth. While this ap-proach has the same effect on the overall CBA as low-ering discount rates, it avoids the problem of distort-ed resource allocation caused by arbitrarily manipulating discount rates.

THE ASSESSMENT FRAMEWORK-A SUMMARY OF FINDINGS

Reviewing the various reports on the issue of forest policy has provided an insight into the process by which decisions are made. This section seeks to summarise the inadequacies that have been identified in the respective assessment frameworks and processes.

Overseas Experience of Assessment Frameworks & Processes

There are several assessment frameworks used in Australia for forest decision-making. Most of the reports have only considered assessment frameworks in general terms, without specific reference to real examples. In order to evaluate these assessment frameworks, the report provides a 'checklist'.

The 'checklist' provides a comprehensive series of questions which any worthwhile assessment framework should be able to answer. The 'checklist' approach facilitates identification of the broad strengths and weaknesses of existing or proposed assessment frameworks

A common theme in all reports (both Australian and international) was an underlying philosophical assumption that a 'technical', non-political choice process was possible, and to some extent desirable. This process was not challenged explicitly in any report, despite the obvious claim that the political arena is the desired place for settling value disputes (indeed it could be argued that this is the raison d'etre of politics). For the moment, however, this discussion accepts the premise that a 'technical' or 'black box' approach is feasible for value choices.

So as to gain an understanding of international experience, the report has sought to review processes suggested by the OECD and the World Bank. These internationally-recommended approaches provide a benchmark for achieving 'world best practice' in this field. The report has also examined the type of system deployed by New Zealand under the auspices of the Resource Management Act (NZ, 1991).

The international reports reflect a worldwide concern over the assessment of environmental values. While it is now recognised that environmental values must be incorporated, it is not known how this should be achieved. The World Bank view is not unusual:

Externalities are thus clearly troublesome, and there is no altogether satisfactory way to deal with them. This is no reason simply to ignore them, however; an attempt should always be made to identify them and, if they appear significant, to measure them. (World Bank, 1992:23)

This accurate, but unhelpful commentary is largely due to a neglect of environmental factors within economic analysis. This was noted by two internationally renowned economists in 1988, with respect to the field of development economics:

Environmental resources appear in this literature about as frequently as rain falls on the Sahara. (Dasgupta and Maler, 1990:102)

They also noted, as recently as 1988, that one of the world's best known texts on development economics (see note below)

... simply has no discussion of environmental resources and their possible bearing on the development processes.' (Das gupta and Maler, 1990:102)

This is symptomatic of neglect within the broad field of economic analysis. The OECD (1992) remarked that conventional cost benefit analysis would not include a number of environmental effects ...because of the lack of market price data or easily accessible direct proxy price data.' (OECD, 1992:49)

Given this neglect the OECD report (1986) on the public management of forests, must be regarded as a landmark publication amongst the reports reviewed in this study.

Overall, given that all economic activity centres on the idea of 'development', the general lack of incorporation of environmental issues is a serious omission. As a result the weaknesses identified in Australian assessment processes are not unusual when compared with 'world best practice'- though this does not make such deficiencies acceptable according to economic theory. Neo-classical economic theory would state quite clearly that on a worldwide basis, as well as locally, such an omission must be regarded as a threat to the economic well-being of society.

Existing Assessment Frameworks

Australia, with its federal political structure, has many different means by which choices over values are made. Rather than describe this diverse array of assessment frameworks, this section will provide a thumbnail sketch of the key features of assessment frameworks and processes. The review of various reports conducted in the previous section, along with the international examples cited, seem to point to three key characteristics of concern in any assessment:

Who Makes the Choices?

'Who makes the decisions' is significant because it determines whose values are brought to the process. The economics profession has long argued that institutions tend to be captured by the values of specific groups, and that to expect any public institution to be value-free and working purely in the public interest is naive. From the perspective of assessment processes, it is important that advocates of differing values are given an equal voice in the process of analysis and decision-making.

What Value Information do they have?

The RAC Forest and Timber Inquiry report took some interest in the Victorian model which involved the Land Conservation Council in decisions about selection of forest logging coupes. The AHC-CALM model of joint survey work also received some favourable attention, though its joint deliberations were designed to identify values rather than make decisions. That no other systems were singled out for favourable comment, in the various reports reviewed in this study, seems to indicate that other processes may be lacking an adequate contribution from representatives of other values.

It is not unfeasible that an appropriately structured institutional process could make optimising decisions-given that the right information on values was available. The information provided, as outlined by the thirteen reports reviewed in this process, is inadequate for the task of making choices over values. The AHC-CALM model received some favourable comments for its process from the ESD and RAC reports Certainly, its assessment method is systematic and thorough. Its prime weakness is that it does not consider, in its own words, 'social values' (AHC, 1992:29). 'Social values' as defined by the AHC, appear to cover that group of values described by economists as existence value. As a result, it only acknowledges some of the values considered important in a total economic assessment. That it was commended as a model for others suggests that even analysis of 'non-social' values is deficient in existing assessment frameworks and processes.

Other reports such as the Victorian Auditor-General's report (AG, 1993) and the NSW Public Accounts Committee

(PAC, 1990) confirm the impression that only poor data is available on values.

Financial values are also a significant part of the analysis. A significant range of reports located deficiencies or made recommendations that pointed to opportunities to improve the gathering of financial data. The deficiencies in this regard extended to the data on both financial values generated by forestry and financial values generated by other activities. The existing assessment framework and processes were not being supplied with the relevant data because the necessary mechanisms such as appropriate accounting systems are not adequate (eg. RAC, 1992:429).

How do they Compare Different Values?

Assessment frameworks must make choices between values. Few technical models were offered on how such choices should be made,outside the political process. The Victorian Auditor-General's report noted that the previous system in that State had used a committee of relevant officials, backed up by a requirement to find options, with a final 'safety valve' being provided by there being recourse to the Minister. No doubt similar methods are used elsewhere, though, what is required is a method that considers all values at the point of decision with appropriate systemic checks and balances.

The AHC-CALM model and the process offered by the RAC Forest and Timber inquiry were the closest Australian examples of technical approaches which are beginning to seek full consideration of all values. The OECD (1986) report is a benchmark in this field. Only the process imrlied in the RAC report comes close to this analysis. It is assumed that if there were another system, the RAC inquiry would have located it. The existing assessment frameworks, obviously, can be said to rely on the judgements of those who take the decisions and the information that they have been supplied with. There must be serious doubts that these existing frameworks have been properly structured. A assessment framework and process needs to specify who decides, with what value information and by what intervalue comparative techniques.

Inadequacies in Assessment Frameworks

The existing assessment framework is in a period of transition. The post-1945 charter of the forest agencies to push for economic development has changed. The capital resources of government and the Australian economy are scarce and need to be deployed for maximum financial benefit or for some clearly identified social 'good'. Changes in social values are also placing pressure on the old system. These two forces, in combination, are leading to greater scrutiny of forest activities, and demanding greater levels of accountability. It is clear that assessment processes are lagging in their response to these issues.

Entrepreneurial Capture

The existing assessment process is vulnerable to 'capture' by entrepreneurs proposing one-o ff developments. This limits government ability to consider a range of project options as proposed by the OECD (1986) and the World Bank (1992) in particular. The existing Australian approach of having a single project option placed in front of decision-makers must be regarded as inferior, if not a structural distortion of the decision-making process.

Lack of Commercially based Accounting of Projects

The lack of commercially-based accounting data for project analysis means that it is not possible to accurately evaluate true project financial worth. This means that many unprofitable activities are slipping through the policy net adding to the economic burden already being borne by Australians. The effect of such methods is to inflate the relative worth of financial values vis a vis non-financial values. It also inflates the relative values of forestry operations vis a vis the alternative activities such as plantations and water production.

Lack of a Comprehensive set of Values Descriptions

The next major flaw in assessment processes is the lack of a complete set of values relating to forestry. Each report examined has mentioned a range of words that seek to describe, in particular, the non-financial values, but there has not been a report which systematically listed what society derives from the forest. It is little wonder then that assessment frameworks barely exist or function if the values are not even identified.

Lack of Focus on the Full Suite of Values

Where non-financial values are considered, they tend to relate only to the assessment of physical characteristics of the forests. All reports acknowledge the need to protect these physical characteristics, for example 'sustainable yield', yet there is evidence that they are not being protected, even at this rudimentary level ((PAC, 1990:55). If physical characteristics are only barely being protected, then it is reasonable to assume that other more abstract and intangible values of the forest are not being taken into account. The Dorrigo Management Area Environmental Impact Statement (FCNSW, 1992:i) does not mention any non-financial values other than soil, birds, animals and recreation. Ecological sustainability and existence values are barely considered at all.

Public Involvement: Preached by Some but not Practised

The practice of assessment seems to place too little stress on the need to obtain public review and input into forestry decisions. In an area where clashes over values are common, the need for representation by different values should be considered. As a result the political process is starved of information about social preferences. The OECD (1986) reveals a considerable commitment to public involvement, in particular their example of the 'Inner Valley Road Project'(OECD, 1986:34).

NFV's are not being Assessed

The most serious flaw is that it appears that assessment of non-financial values is simply not occurring. The concentration of reports on discussing aspects of how such analysis could be done, rather than a focus on how it is being done leads to the conclusion that serious analysis of the relative worth of non-financial values is yet to begin. It would seem fair comment that if relatively simple choices between wood and water production are not being evaluated (eg. Read and Sturgess, 1992), then other more complex interactions are also being neglected.

Future Generations Considered by Implication Only

The OECD (1986) report was not without its flaws. Its assessment framework gave little attention to considering the impact of future generations from any valuation assessment process. In economic theory it is impossible to internalise the interests of future generations simply by assessing the preferences of the present generation. Neither market processes or valuation techniques, in the face of irreversibility, are capable of 'looking after' the interest of future generations in any manner that can be relied upon. Fortunately the concept of ecological sustainability goes some way to achieving this goal. The internalisation of the interests of future generations requires explicit discussion.

Most reports steered clear of identifying procedures that would resolve the choices thrown up in the forestry debate. The RAC report (1992) went furthest in this regard but tended to educate by demonstration rather than explicit prescription. The ESD report (1991) had some brief comments, but if the assessment of these different values is to proceed satisfactorily, it is clear that the hard issue of 'assessment frameworks' must be dealt with honestly.

The OECD (1986 and 1992) indicated that other nations are moving towards direct evaluations of non-financial values. Australia's approach would appear to be somewhat slower than average, though of course there are some institutions such as the World Bank that have acknowledged the need for change but are yet to make the practical changes that are required. It was significant that the United States was acknowledged as having greater experience in the area of non-monetary evaluation. The experience was deemed to be due to the 1969 Environmental Policy Act which required quantification in non-monetary terms of all environmental impacts. The OECD concluded that it would seem that education and more valuation projects are a prerequisite for more effective utilisation of the techniques. A learning process is needed. (OECD, 1992:38).

Valuation techniques, whether monetary or non-monetary, form the backbone of an assessment framework. The same learning process is required for assessment processes as well. Many OECD countries are learning but Australia does not appear to have started.

Assessing the Assessment Frameworks

A major difficulty with evaluating an assessment framework is the sheer scale of the task itself, as anyone who has read an Environmental Impact Statement will testify. The result is a tendency to focus on one or two key angles to the exclusion of other angles. Table 6 seeks to provide a systematic 'framework' for assessing assessment frameworks. A series of questions is provided, that an assessment framework, if it is to be successful, should be able to address. The existing Australian and international assessment practices are evaluated The table has been tested on actual reports, in this case the Victorian Auditor General's report (AG, 1993) as well as an analysis of the Dorrigo Management Area Environmental Impact Statement (FCNSW, 1992).

The native forest is capable of producing three major categories of values. These are:

- financial values from timber production;
- alternative financial values from other forms of forest use; and
- non-financial values.

This report has sought to examine the issue of giving each type of value 'full and due consideration' in policy processes. Three themes have recurred throughout this study.

- There is a theme focusing on the analysis of technical comparisons between different values. That is the use and abuse of various techniques that quantify values.

The native forest is capable of producing three major categories of values. These are:

- financial values from timber production;
- alternative financial values from other forms of forest use; and
- non-financial values.

This report has sought to examine the issue of giving each type of value 'full and due consideration' in policy processes. Three themes have recurred throughout this study.

- There is a theme focusing on the analysis of technical comparisons between different values. That is the use and abuse of various techniques that quantify values.

- Asecond theme revolves around the nature of the institutional arrangements that ensure value selection which is socially optimal.
- The major theme, however, is achieving a level playing field between the major types of values derived from the forest.

The emphasis on achieving a level playing field is in the spirit of the microeconomic reform process that has taken place over recent years, and continues with recent reports such as the Hilmer Inquiry (NCP, 1993). Sims described microeconomic reform as occurring when changes are made to achieve more output from a given level of inputs. (Galligan et al, 1993:15).

The output from the forests is intended to maximise the social welfare of the community. For this reason, microeconomic reform in the forest sector must seek to arrange the three sets of values above so that welfare is maximised.

The implication of the need for micro-economic reform is that the sector is not making an optimal contribution to the quality of Australian life. This review of reports on the forest sector suggest that there is considerable potential for microeconomic reform in two dimensions.

- In broad social terms it would appear that the community is not maximising the social benefit from the forest resource.
- Secondly, within the social benefit the community receives, the financial value to be had from the forest resource is not being maximised, assuming that all other types of values remain unchanged (ceteris paribus 'All other things being equal.').

If the microeconomic reform process (optimising social benefits) in the forest sector is to succeed, the estimation of the worth of the three categories of value described above must be accurate. It would appear that estimation of these three categories of value is presently inaccurate. This inaccuracy would confound any attempt to derive optimal social benefits.

Effects of Miscalculating Financial Values

The impact of miscalculating the financial value of timber production is significant for Australia's macro-economic performance. Firstly, it overestimates the financial worth of timber production vis a vis some other form of commercial activity. The Read Sturgess report (1992) made it clear that it was more financially valuable for the Upper Thomson catchment to supply water than to supply timber, and consequently more profitable for the economy. The overestimation of the worth of forestry would tend to hide this potential economic 'free lunch'. Other competing uses for the forest may be similarly affected. More generally, the over-valuation of forestry financial values disguises what may be net losses in this sector which are actually diminishing the wealth of Australia unlike other commercially viable industries.

Failure to accurately value forestry also leaves the commercial operations of the forestry commissions vulnerable to an income shortfall in the event of corporatisation. Even without corporatisation, the poor measurement processes for forestry blind the commissions to other profit centres such as leasing forests to supply water to regional communities with expanding industries.

Over-valuation of forestry through poor input pricing, in addition to affecting competing uses of the forest, also undermines industries that provide substitutes for forest timber ranging from plantations to steel production. Inefficient input pricing in native timber production undermines all competing industries, all other things being equal. This report has focused on the 'collateral damage' caused to the plantation industry. However, what applies to the private plantation industry, also applies, probably on a lesser scale, to other industries that compete with native forest timber. The private plantation industry (NPAC, 1991) is one clearly visible victim of an 'unlevel' playing field in the forest sector.

The over-valuation of native forest harvesting undervalues, by definition, the product of plantations. This leads to lower levels of investment, employment and wealth creation than may otherwise have occurred. It prevents the development of value-added timber industries that can further develop on the basis of a plantation resource. This presents the Australian economy with a 'double loss'. T h e macro-economy loses, marginally at least, from harvesting native forests, and then loses again through the disincentive effects on plantations. Other industries may suffer as well. Finally, there is a third loss built into the playing field, which are the microeconomic distortions, such as poor taxation arrangements, identified by NPAC (1991), that inhibit the plantation industry. The microeconomic distortions in the plantation sector alter the financial calculus such that the value of native timber is enhanced relative to plantation timber. In addition these microeconomic distortions cause a further round of macro-economic losses from the plantation industry.

The structural distortions within the forestry sector, and within other sectors both competing and substitute, compound the overvaluation of forest timber relative to other values. This causes a loss of income, jobs and a waste of capital as private agents respond to these inappropriate signals.

The Effects of Miscalculating NFV's

The second major impact of overvaluing forestry financial values, is felt on non-financial values. Over-estimating of forestry financial values also underestimates, by definition, the relative worth of non-financial values. Thus the process of evaluating non-financial values is seriously handicapped from the beginning. The significance of this was pointed out by Sims, in another context, when he stated

Clearly our objectives are broader than maximising GDP. For example pollution in our cities is of general concern. It might be that including externalities in prices would decrease environmental damage and decrease growth, but improve overall living standards. (Galligan et al, 1993:17, emphasis added)

Two points flow from the comments by Sims.

Firstly, Sims appears to be giving emphasis to a commonly-held view that environmental improvement can only be achieved at a cost to the economy. This review has indicated, in a very clear manner, that in the forest sector financial gains (see note below) can be achieved whilst making environmental improvements. The view espoused by Sims presumes that the 'balance' between environmental and financial considerations is 'efficient'; that is, in the context of this report, the evaluation of the financial worth of native forests has been accurately calculated. The evidence indicates that the balance is inefficient, because calculations of financial and alternative financial values, as well as NFV's are inaccurate. Therefore, this report indicates that improvement in environmental conditions, over a very large margin, will improve financial conditions as well.

At some point along the policy continuum, improving environmental conditions will negatively affect financial conditions-but the forest sector, along with many other sectors, is not at this point, and indeed is a long way from it. Microeconomic reform in the forest sector has the potential to bring environmental improvement in its wake by removing financial inefficiencies that have contributed to over-exploitation of the environment. The implication of financial 'inefficiency' in the sector needs to be clearly understood. It implies that society could be financially better off if certain microeconomic reforms were undertaken. The implication of 'economic' inefficiency is that, in the context of this report, the environment and the financial wealth of society, could be better off, leading to an unambiguous improvement in society's level of 'satisfaction'. In this sense, when 'financial inefficiency' is present, microcconomic reform and environmental policy can be complementary, and not competing policies.

Secondly, failure to evaluate non-financial values relative to financial values accurately is critically important to maximising 'overall living standards'. The benefits that are provided to Australians from NFV's can also be described as 'psychic income' (see note below).

These benefits also have the fortuitous 'collateral' benefit of also attracting tourists, and hence contributing to the expansion of this critical export industry.

Non-financial values provide direct inputs into the daily consumption of Australian citizens and even to foreigners through tourism, wildlife films and existence values. An assessment framework which seeks to optimise benefits to the community, but which does not recognise these non-financial values can generate financial flows that actually undermine community well-being.

This happens because the source of much 'psychic income' (ie. the natural environment) has been lost and replaced by financial benefits that are too small to compensate. If non-financial values are to be given their full and due consideration, then financial and alternative financial values must be correctly evaluated.

This review has identified that there are significant inefficiencies in the estimation of value in the forest sector.

The presence of these inefficiencies in the market activities of the forest sector is due to structural distortions.

The structural distortions contribute to poor financial and environmental performance. There is a considerable social, environmental and financial dividend to be achieved by microeconomic reform in the forest sector. Microeconomic reform requires that a 'trilogy' of assessments be made: evaluate financial values (FV's), evaluate alternative financial values (AFV's) and evaluate non-financial values (NFV 's). Continuation of present practices such as failure to 'estimate value' accurately will cost Australia in lost jobs, lower incomes and a diminished environmental quality of life.

Here this report distinguishes between financial gains that contribute to the overall health of the financial economy, and economic gains that contribute to the overall state of social well-being. Improvements in financial well-being are commonly referred to as economic gains, for example improved resource allocation, because of the easy but methodologically sloppy habit of referring to the financial activities of the Australian society as the 'economy'. Accurately the economy is the process by which we gain 'welfare', a part of which is the financial gains offered by financial activity.

In another circumstance, when a Treasury officer was asked to described the non-pecuniary benefits of working in Treasury, he called these benefits 'psychic income'.

2

ROLE OF FORESTRY AND TREES IN THE NATIONAL ECONOMY

INTRODUCTION

As a small open market, forestry and related activities can play only a supplementary role in the Antigua and Barbuda economy which is primarily driven by the tourism industry, in the wake of the steady decline of agriculture.

Between 1977 and 1990, the country experienced strong economic growth, supporting one of the highest per capita standards of living among the OECS countries in the Eastern Caribbean. During this period, real GDP increased from EC $158.8 millions to EC $355.9 millions, an average growth rate of 6.2% per annum. However, the Government has been running a fiscal deficit for a number of years, having financed its capital investment programme through loans, a significant portion of which are denominated in foreign currencies. The Government has therefore announced austerity measures which include a freeze on public sector spending, more effective revenue collection and divestment of assets.

Nevertheless, tourism continues to grow, contributing an estimated 60% to the economy. However, it has placed considerable demands on the local infrastructure, particularly water and sewage, and crowded out alternative investment opportunities, while severely stressing the use of the natural environment for construction of additional resort facilities. The Forest Economics consultant has pointed out that changes in the market place can be affected by changes in societal values and a variety of products in the developed countries have suffered because they were determined as being "environmentally unfriendly". The tourist industry is dependent on visitors from the metropolitan countries where this discrimination is particularly rife. In these circumstances, it would be prudent for the Government to ensure that natural resources which enhance the tourism plant are well protected and, at the same time, properly utilised.

In this respect, the GOAB needs to adopt an enlightened approach to the planning, development and use of land and to ensure that such a policy is strictly observed. Under such a policy, forestry's role should be recognised as being essential to the protection and conservation of critical resources, such as water, and complementary to developing the tourism product, as well as the quality of life of the people of Antigua and Barbuda.

Other benefits which can accrue from forestry in Antigua and Barbuda include:

(i) rural employment opportunities;

(ii) shelter from the desiccating effects of the prevailing winds;

(iii) land stabilisation and soil conservation/amelioration;

(iv) production of fence posts, forage for livestock, fuelwood, fruits, nuts and other products;

(v) protection of natural habitats and wildlife for recreation, scientific research and tourism.

Despite the reputed benefits of the tourism industry, the Government should concentrate its attention on all opportunities for economic growth and optimal use of natural resources. In tandem with other national strategies, the role of forestry and trees needs to be heightened to become an integral contributor in maintaining living standards on a sustained basis.

RELEVANT DEVELOPMENT POLICIES

In pursuing the development of the tourist sector, the GOAB appears to have neglected agriculture and the need to maintain stability in all sectors. Indeed, there seems to have been an absence of clear Government policies to guide the society's growth and improvement. However, it is pleasing to report that within the last two years a Ministry of Planning has been established in Antigua & Barbuda with the responsibility for developing a National Development Plan.

This new Ministry has initiated an integrated planning process which will take some time to fill the existing void. Such planning will take account of the country's present economic circumstances, the nature of its physical resources, its human resource capabilities and requirements in order to determine fundamental national strategies on which future growth can rely. In the short-term, the austerity measures announced must be equitably implemented as soon as possible, along with appropriate measures to ensure cohesion among the various sectors. The proposed NFAP recognizes these needs, and its multi-disciplinary approach is recommended for adoption by the Government in planning for the country's future.

Legislation

Antigua and Barbuda's forest legislation (The Forestry Act, Cap. 99) dates back to 1941 and provides for, inter alia, the establishment of reserves, exploitation permits, clearing and afforestation on private lands, etc. The Forest Regulations deal with permits, control of fires, cattle trespass and survey and demarcation of Forest Reserve boundaries. Unfortunately, the Act and Regulations are no longer enforced, and uncontrolled exploitation (mainly wattle for fish and lobster traps) has been practised. However, the forest cover is still relatively intact, but species diversity has been impaired.

The Legislation consultant has discussed related legislation and described various deficiencies existing in the enactments supporting resource and environmental management. Related legislation includes:

- The Land Development and Control Act (1977) for physical development planning;
- The Fisheries Act (1983) for protection of mangroves and habitats of fauna and flora;
- The National Parks Act (1984) to protect and develop ecological resources;
- Antigua Agricultural Development Corporation Act (1978) to stimulate agriculture (including forestry) and undertake commercial management.

The deficiencies identified underscore the need for new legislation in the forestry and water resources areas, together with amendments to other legislation, especially the National Parks Act, as well as the need for regulations to provide guidance for environmental managers. Amendment to the National Parks Act should include consolidation of existing laws relating to coastal zone management and declaration of protected areas, viz. Beach Protection Act (1957), Beach Control Act (1959), Marine Areas (Preservation and Enhancement) Act (1972), Public Parks Ordinance and the Botanic Gardens Ordinancc.

New and amended legislation should provide for: structured integrated environmental planning and control; statutorily determined time-frames for completion and approval of national and sectoral plans; and adequate financial and legal incentives for sound agroforestry development in the country. This latter element is required to facilitate agricultural credit and loan facilities to encourage farmers. The consultant also advocates umbrella environmental legislation as is found in the Bahamas, Jamaica and St. Kitts and Nevis, and which is actively being considered in Guyana and Trinidad and Tobago. Such legislation should empower the Historical, Conservation and Environmental Commission as a statutory body with the necessary functions in this regard.

Many of the statutes cover the same areas, while many issues are not covered by any existing legislation. As in other CARICOM countries, this situation has arisen because legal development has tended to deal with single issues, due to perceived necessities or pressing concerns. This patchwork of fragmented and uncoordinated legislation is not rooted in any central theme and has contributed to confusion among agencies, non-enforcement of regulations and a dilution of responsibility. It is hoped that the GOAB will be committed to halting environmental degradation and to improving the quality of life for its citizens as well as the quality of enjoyment for its visitors. It should therefore move urgently towards rationalization of environmental legislation to ensure adequate provisions for protection and development of forests, natural habitats and wildlife in the interest of conservation of the environment and biological diversity.

Administration

Forestry falls under the portfolio of the Ministry of Agriculture, Fisheries, Lands and Housing and is staffed by a professional officer, two sub-professionals four forest rangers, a nursery technician and a trainee. This is notable growth for the country which has never had the benefit of such capacity, even in colonial days.

The GOAB now needs to structure this team into a proper Forestry Division by providing it with clear functions and responsibilities and with adequate facilities to perform these duties effectively. This will require the promulgation of an official forest policy which, until now, has been notably lacking.

CONSTRAINTS AFFECTING THE SUBSECTOR'S DEVELOPMENT

As emphasized in many reports during the past decade, the main constraint is the lack of a coherent land use policy to guide the allocation of land for suitable uses and to provide security of tenure for those in dire need of adequate land space. Consequently, there has been no survey nor gazetting of forest reserves for fifty years since the Forestry Act and hence, there is no forest estate in Antigua and Barbuda. The absence of a land use policy has prevented the development of afforestation schemes and will be a serious hindrance to promoting agroforestry in the country, if not addressed urgently. At the same time, there is no forest policy nor the financial means to support one. However, the Government has moved steadily to provide trained personnel, although it has taken many years.

A potential constraint to the implementation of the NFAP could be the profusion of overlapping jurisdictions caused by the existing enactments. However, the dialogue initiated by the Historical, Conservation and Environmental Commission and the apparent coordinated support engendered by the TFAP suggest that suitable rationalization of these overlaps can be achieved.

The exercise was carried out with the close collaboration and active participation of national counterparts from relevant public agencies and representatives of the private sector and NGOs. Working papers were prepared for circulation and for use as reference documents. These papers were used by the author in synthesising the main report in which the situation of Antigua and Barbuda was analyzed according to the five TFAP priority areas (see regional overview document). Based on these analyses, and group discussions of the major issues during the mission, strategies and action programmes were developed to attract technical assistance and investment with the aim of optimising the contribution which forestry and tree cover can make to sustainable social and economic development of the country.

The report, therefore, represents a National Forestry Action Plan (NFAP) for Antigua and Barbuda. The NFAP recognises the limitations of the country's economy but stresses the long-term benefits of environmental management and development. The Government endorsed the proposals before they were presented to the public for discussion at a National Round Table held in June, 1991. The active and meaningful feedback generated allowed the draft NFAP document to be strengthened and made available to the international community for consideration of technical and financial support. It is hoped that this International Round Table will be convened towards the latter half of 1993.

Reference should also be made to related documents:

Regional Overview of the CARICOM/FAO Tropical Forests Action Programme in the East Caribbean. CARICOM/ FAO/BDDC-ODA, July 1993.

Forestry development initiatives at the regional/ subregional level-regional project proposals 01-05.

The Status of Mangrove Conservation in the CARICOM Islands of the Eastern Caribbean. Report to the Commission of the European Communities as part of the TFAP for the Caribbean Region. P. Bacon, 1991.

Natural Regions and Ecosystems

Antigua and Barbuda is located between latitude 17°0' and 17°45' N and longitude 62040' and 62°54' W, and has a total land area of 440 km 2 (Antigua 280 km2, Barbuda

160 km2) and population of 65,200 (Antigua 64,000, Barbuda 1,200).

The climate is tropical-maritime, with a mean annual temperature of 26°C, varying from 16.7 to 28.3°C. Average annual precipitation is 1100 mm, but variable throughout the year; the islands are susceptible to severe droughts (as well as earthquakes and hurricanes). The main rainy season is from October to November.

The geology of Antigua consists of a hilly limestone zone to the east, a central plain of marl and clay overlying volcanic material, and volcanic mountains in the southwest rising to 410 m above sea level at Boggy Peak. The topography is mainly gently undulating with slopes of 11-20 degrees. Shallow soils overly volcanic material and the limestone, over most of the island. Barbuda, which lies 32 miles north west of Antigua, is a low-lying coral limestone island, no higher than 60 m in the east, but with a large coastal lagoon enclosed by a long sand spit in the west.

Like most of the Caribbean Islands, Antigua was completely wooded when "discovered", but following European settlement it was almost totally cleared to make way for cultivation of sugar cane and cotton. With sugar as the dominant crop for over 300 years, Antigua & Barbuda's natural landscape was undergoing irreversible changes which created negative pressures on wildlife habitat and watershed areas. The demise of the sugar industry in the 1970's led to the abandonment of sugar cane plantations. Presently, Barbuda is affected by the overgrazing of animals which have continually removed valuable vegetation protecting watersheds.

As described by the Wildlife/National Parks consultant, the ecosystems of Antigua and Barbuda range from evergreen woodlands, xerophytic woodlands, scrublands, grasslands to mangrove forests, herbaceous swamps, saltponds, sandy beaches, rocky shores, coastal lagoons, sea grass beds, coral reefs, coral reef outcrops and oceanic islands and rocks. Ecosystem variety is enhanced by the presence of caves in the north-eastern limestone area and by natural seasonal drainage channels, streams, potholes and ponds. Barbuda's giant coastal lagoon is unique to the Region and together with its extensive tidal flats, sand bars, underwater sand dunes, salt ponds, cliffs, caves and "highlands", all providing special habitats for wildlife. Wildlife species associated with these natural ecosystems have developed physical, behavioural and physiological adaptations.

REVIEW OF THE FORESTRY SECTOR

With the decline in agriculture, the area of land under tree cover has increased over the years, and various reports estimate that there are 9,254 ha of woodland (33% of Antigua's land area) and 10,000 ha of scrub growth (36% of the land area). Table below classifies the woodland areas of Antigua.

Table : Woodland Composition for Antigua

WOODLAND	AREA (hectares)	WOODED AREA (percentage)
Moist Forest	2,200	24
Dry Woodland	5,654	61
Cactus Scrub	1,100	12
Mangroves	300	3
TOTAL	9,254	100

The species composition and structure of the vegetation has been described (Beard, 1949) and although all the forests are of secondary growth, they contain a number of valuable timber species, including mahogany (Swietenia mahagoni) and white cedar (Tabebuia pallida). On the present sites, however, they are unlikely to produce quality timber, as most of the forested lands are only suitable for conservation purposes.

Table : Woodland Distribution for Barbuda.

CATEGORY	AREA (hectares)	LAND AREA (percentage)
Wooded Areas	10,619	66
Mangrove Swamp	3,533	22
Scrub & Rough Grazing	1,285	8

FORESTRY IN LAND USE

Land Use Distribution: Present land use distribution is difficult to quantify reliably, as various estimates differ. The Land Use/Watershed Management consultant has summarized previous percentage estimates as shown in Table below.

Table : Antigua & Barbuda Land Use Distribution.

CATEGORY	ANTIGUA LAND AREA (%)	BARBUDA LAND AREA (%)
Woodland	34	66
Mangrove Swamp	3	22
Scrub/Rough Grazing	33	9
Improved Pasture	4	-
Agriculture Crops	8	1
Settlements	11	-
Tourism	3	2
Airfields	3	-
Reservoirs	1	-

Rural land use activities include beekeeping, livestock, small cropping, large specialty farms and part-time farms. The main crops are pineapples, cotton, onions, pawpaw, coconuts and flowers. There is considerable opportunity to increase local production of many tropical fruits and to reduce imports and outflows of foreign exchange. This strategy can be integrated with a livestock improvement programme to reduce imports of meats. The feed for livestock would be enhanced by fodder and crop residues, while the animal litter would enrich the soils and improve their productivity. Numerous reports on land use have pointed to the need for watershed rehabilitation to enhance the ground water storage, reduce wind desiccation, increase fodder and fencing material. NGOs have become active in advocating tree planting for food production and amenity of the countryside, employing women and community groups as well as school children. In this regard, the Pan American Development Foundation and the Environmental Awareness Group have embarked on a small scale agroforestry project, working with about twelve communities to establish fruit orchards, live fencing and woodlots.

In Antigua the Government owns sixty percent of the land, some of which is currently rented to small farmers on a year-to-year basis. This system is practically feudalistic and offers no stability to the farmers who need to feel that they have a stake in their nation. It should be replaced with longer term leases and also by a system of freehold through a land settlement scheme based on economically sized parcels. Indeed, GOAB should consider divesting its farms and stations to the highest bidder, or to staff and workers in part settlement for voluntary redundancy. This would relieve pressure on the Recurrent Budget, while allowing inflows of cash to support necessary investments or loan repayments.

Watershed Management

Antigua and Barbuda depends mainly on surface runoff for its water needs: domestic, industrial, irrigation and recreational. Existing ground water resources are extremely limited, and reportedly are threatened by excessive use, as well as potential contamination from salt intrusion, chemical pollutants and inadequate sewage and waste disposal facilities. While efforts are in train to cope with this problems, attention needs to be given to measures that will reduce siltation of reservoirs and maintain the land in a conducive condition for optimal percolation and soil water retention.

There is very little data on surface runoff and infiltration, but over the years, the change in vegetation has obviously affected the volume of runoff and the transport of sediment into the reservoirs and ponds. Additional water sources include wells, the desalinization of sea water and cisterns.

Total demand is 3-4 mgd (million gallons per day) of which actual consumption was 2.42 mgd in 1987. Delivery capacity is 2.2-3.8 mgd, there being considerable loss from broken pipes. Future demand is estimated at 5.1 mgd in the year 2000 and approximately 8.0 mgd in 2020. A proper water plan is needed particularly in view of the concentration on tourism and the increasing visits of cruise ships which combined, can take away from Antigua approximately 500,000 to 750,000 gallons of water per month.

While interest in watershed areas is primarily focused on water quality and quantity for storage sites, this may shift with increasing demand for food, to include crop production along with systems to minimize erosion and pollution. Roaming livestock have compacted the hillside lands and have browsed on cultivations and gardens.

As a possible solution for such a problem, it is recommended that agroforestry (multi-cropping) systems be used (eg. plant leucaena in contour hedges with cultivated grasses). Also, the regular illicit burning of rough grazing and woodland causes the spread of citronella grass and creates a greater problem for land rehabilitation.

Before embarking on any watershed programme, a detailed plan of the area should be developed, including the location of agroforestry treatments, conservation measures, subdivision of farm holdings, protection of runoff and drainage channels.

Agroforestry

Within the framework of integrated land use planning and watershed management, the production of agricultural crops in combination with trees is an excellent conservation practice which offers the opportunity for sustained use of land to maximize returns, while protecting soil and water resources. In Antigua, agroforestry is not new, having been practised historically on large and small holdings. However, there is room for intelligent modifications to be applied according to local situations, and the approach is being proposed as a means of increasing land utilization and farm output, particularly fruits and livestock, primarily for the local market. In addition, meaningful proposals have been made for solving the problem of uncontrolled grazing, while providing forage, fencing, shelter and pasturage for livestock.

Forest Management and Silviculture

The present forest cover of Antigua is secondary forest, which has developed following the decline of the sugar industry. The most important species are Spanish oak (Inga laurina), silk cotton (Ceiba pentandra), mahogany, turpentine (Bursera simaruba), white cedar, and logwood (Haematoxylon campechianum). The mangrove forests consist of Rhizophora mangle. Avicennia nitida and Laguncularia racemosa. The dry scrub forests are dominated by Acacia spp. and Prosopis chilensis which were introduced and have become naturalized. Other introduced species include casuarina, neem (Azadirachta indica), tamarind (Tamarindus indica), leucaena (Leucaena leucocephala) and woman's tongue (Albizzia lebbeck).

It is not expected that the secondary forest will contain many trees of marketable size, and indeed there is no sawmilling capacity on the island. However, on the few plots of planted trees (neem, casuarina, leucaena, mahogany and white cedar), the growth has been promising, and annual measurement of selected stems on each site can be made by the Division's staff on a systematic basis. The data derived will assist in future determinations of species suitability and potential performance, should a large reforestation programme be adopted.

The Forestry Division can now be assigned responsibility for the management of Antigua's forest estate (when demarcated and gazetted) and for implementing a forestry programme based on the recommendations of previous studies and the NFAP. Existing forest cover and scrubland should be assessed in terms of the opportunity to conserve biodiversity or, where possible because of soil depth, to be reforested for the production of commercial timber and other products. This will entail air photo interpretation ground-truthing as a normal Divisional function, and preparation of maps and working plans. Forest rangers should be deployed in the field on regular patrols against illicit extraction, cattle trespass, clearing, fires, etc.

Forest Based Industrial Development

There is no industrial forestry activity apart from small scale charcoal production (largely for tourist and domestic recreation pursuits), extraction of wattle for fish and lobster traps, and some use of mahogany and white cedar for boat construction and fence posts. Estimates are that 1135 tonnes/year of charcoal and 4000 fence posts/year are produced, but there are no supporting data. The wood related manufacturing sector is also small, but again there are no records. Some opportunity may exist for handicraft material for carving, but technical training would be required to enhance growth in this area, particularly to utilize the limbs pruned from urban mahogany trees, some of which are dying or so badly mutilated that they should be felled and replaced. In addition, there is potential for non-wood products of agroforestry such as fruits and nuts, fodder and water.

Neem has been quite successful in Antigua and Barbuda and is widespread, mainly along road-sides or fence lines. It seems to have withstood the effects of hurricane Hugo in 1989, and attention should be given to determining the utilization prospects for this valuable fast-growing species. It is being planted elsewhere in the world and produces a wide range of products.

These include the extract from the seed kernel and seed coat to yield neem oil, which is used as a lubricant, insect repellant and insecticide and in the manufacture of soaps and cosmetics. The seedcake residue is a valuable organic manure and cattle feed, while the root exudate has medicinal properties; the bark yields an industrial gum and the leaves are used as a green manure. The timber is also useful for lumber and fuelwood. Antigua and Barbuda has an excellent opportunity to investigate the industrial potential of this tree which could attract investment to the country.

Fuelwood and Energy

In spite of the relatively high per capita income levels in Antigua and Barbuda, there is still some consumption of charcoal. Unfortunately, no reliable records exist to indicate the levels of production and consumption. Charcoal burning does not presently constitute a threat to forest resources as suitable material is already available from pastures in the form of Acacia.

This tree is a pest which threatens the health of cattle, as the thorns cause severe skin lesions, exposing the animals to the risk of disease. The GOAB is obviously aware of the continuing world energy crisis, and a case could therefore be made for encouraging the use of alternate sources of energy to save foreign exchange and to utilize natural products generated in the course of development initiatives such as agroforestry and watershed rehabilitation.

CONSERVATION OF FOREST

In spite of its small size, the island possesses considerable wildlife and ecosystem diversity, much of which could enhance the growth of tourism but, more importantly, could awaken society's consciousness and pride in its natural environment. The Country Environmental Profile has recommended specific elements to be included in a system of national parks and protected areas, and these should be expanded to include coastal zones and forested areas.

Negative development pressure on the coastal zone can compromise the future ability of the country to sustain long-term socio-economic growth. All development proposals should be carefully reassessed to avoid adverse environmental impact. In the meantime, interim action can be taken to stimulate financing of resource management through collection of fees by local guides and monitoring of visitor levels and impact. Resource inventories are proposed under the NFAP projects.

Institutions

The main institutions concerned with forestry and environmental management include the Forestry and Fisheries Divisions, the Development Control Authority (DCA), the National Parks Authority (NPA) and NGOs, primarily the Environmental Awareness Group (EAG). Under the present tight fiscal policies, the public agencies are restricted in their ability to execute programmes. There are no funds available to undertake work programmes in forestry or agroforestry, to maintain monitoring of fisheries and fishermen, nor to expand the capability of the NPA. The NGOs led by the EAG, are particularly knowledgeable and committed, and though lacking in resources, they are quite resourceful. They have played a significant role in raising environmental consciousness, although their confrontational approach has been met with official suspicion. However, they are now widely accepted in official circles that recognize the desirability of closer collaboration.

This has led to the appointment of the Historical, Conservation and Environmental Commission (HCEC), aimed at improved dialogue and coordination within the public sector. However, the general institutional situation of the public sector is weak in implementing capacity and performance, and the NGOs are still skeptical that the Government will change its ways.

The NPA is backed by strong legislation which provides for inter-sectoral linkages to consult and cooperate with agencies having similar aims and objectives. It also has the power to delegate its functions and this could be done through the Barbuda Council, Forestry and Fisheries Divisions, the EAG and other NGOs.

The Barbuda Council was established by the Barbuda Local Government Act (No. 15, 1976), to manage Barbuda's government affairs. The Council consists of 9 members who elect a chairman and secretary from among them. Proposals for capital and special expenditure are required to be submitted to the Ministry of Home Affairs (Central Government) who forward them to the Ministry of Finance. The Council has powers to make bye-laws for development on the Island.

The conservation and development of forest resources can only be implemented by capable resource management institutions, properly staffed and financed, and provided with the necessary authority for efficient execution of their responsibilities. Unfortunately, the lead agencies in Antigua and Barbuda do not seem to be adequately equipped at present, and will require considerable strengthening to improve their management capacity. With the recent return of the country's first professional forester, the Government now possesses the nucleus of trained forestry personnel. The Forestry Division can now be empowered with the following responsibilities and functions:

(a) protection of the forest estate, wildlife sanctuaries and protected areas;

(b) afforestation and forest management in watershed areas;

(c) cooperation with private and Government agencies on urban forestry programmes (including the Botanical Gardens), forestry education and information;

(d) assessment and conservation of biodiversity and promotion of ecotourism;

(e) generation of self sustaining activities utilizing forest and non-wood products;

(f) preparation of budget plans and annual performance reports.

The following steps are therefore recommended for official approval and action:

(i) adoption of a formal national forest policy and appointment of the Chief Forestry and Wildlife Officer;

(ii) declaration and gazetting of the forest estate, followed by survey and demarcation of forest reserve boundaries;

(iii) preparation and approval of a baseline work plan for the Division, including job descriptions for all staff;

(iv) transfer of game wardens, caretakers and necessary staff to the budget of the Division;

(v) assignment of responsibility for supervising the maintenance of the Victoria Park and Botanical Gardens;

(vi) assignment of office accommodation, vehicles and equipment to enable efficient performance of the Divisions functions.

After establishment of the Division, there are a number of basic activities which it can undertake, namely:

- planning, layout and implementation of a forest recreation centre at Wallings;
- planning and implementation of urban forestry based on the Botanical Gardens;
- field assessment of planted forested trees, enumeration, diameter and height measurements, site conditions;
- assessment of charcoal production from leucaena at Dunbars;
- continued cooperation with Fisheries Division and NGOs in mangrove and wildlife monitoring;
- initiation of contacts with NPA to discuss areas of collaboration;
- selection of possible reforestation sites and planning proposals for employment absorption.

These activities can be undertaken with little or no new financing and are within the technical capability of the trained staff. The results will soon be evident if tasks are approached in a business-like manner and with an eye to the future.

PLAN OF ACTION

Based on the results of the Country Mission, as outlined in the previous sections of this report, and on the working documents prepared by various consultants, a Plan of Action is proposed for the development of the country's forestry subsector. The National Forestry Action Plan (NFAP) proposes elements of national development policies, a long-term policy for the sub-sector, together with priority area policies and objectives on which specific action projects are proposed for a five-year period.

Mindful of the country's present financial situation, the Plan proposes redeployment of the existing personnel and budget as the Government's initial investment in the sub-sector to attract donor support in setting a sound foundation for sustainability. The plan includes the normal work of the Forestry Division which will also function as lead agency in the implementation of three projects.

The NFAP is estimated to cost a total of US $3.0 million of which US $2.2 million would be contributed by international donors while the remaining US $0.8 million would be contributed by Antigua and Barbuda, mainly in kind.

National Development Policies

The adoption of national development policies for land use planning, land allocation and sustainable resources management are recommended for enhancing the contribution which forestry and trees can make to socio-economic development and environmental stability.

Long-term Subsector Policy and Objectives

59. It is proposed that the long-term Forest Policy of Antigua and Barbuda should be:

(a) to protect the country's natural forest resources;

(b) to promote the development of forests and tree cover on suitable lands;

(c) to encourage the application of agroforestry in planned land use for rural development;

(d) to pursue opportunities for utilization of forest and other products.

In pursuing this forest policy, the Government will seek to ensure:

- the wise use and proper management of the island's limited forest and natural resources;
- the preservation of representative samples of the island's natural forest and vegetation resources and the maintenance of the ecological balance;
- the provision of opportunities for the public for scientific, educational and recreational activities to enhance knowledge of forest ecosystems, research in potential utilization of the gene pool and tourism benefits;
- the stimulation of industries capable of boosting the island's economy, based on products derived from forests and trees.

Therefore, the proposed objective of the NFAP is to establish and manage a national forest estate on Government and private lands. This will ensure conservation of soil and water resources, protection of forest ecosystems and production of wood and related products for social and economic benefits on a sustainable basis.

Priority Area Policies and Projects

In the medium term, the NFAP will be guided by the priority area policies and strategies outlined below in the implementation of the respective action projects. These policies are all inter-related and represent the minimum actions required to effect the fundamental changes necessary towards achieving the basic aims of the new long-term forest policy.

Institutions-to provide effective environmental legislation and capable institutions for proper administration and sustained development of forestry and related programmes, with the participation of local people and interest groups.

The purpose of this policy is to improve the environmental management-capacity of the country and to install multi-sectoral and participatory mechanisms for planning and monitoring of forestry and related programs. This aspect is of the highest priority if the national long-term policy and objective is to be attained. Accordingly, the following project is proposed for immediate attention:

Country Capacity for NFAP Implementation

The objective of the project is to build capacity in the lead agencies to ensure initiation and co-ordination of NFAP implementation, sustainable natural resources administration and management, project preparation and management, public education and extension, and environmental management for conservation of natural resources.

The project includes specific provision for the survey and demarcation of forest reserve boundaries in order to create a forest estate.

Forestry in Land Use-the proposed policy recognizes the multiple benefits which accrue from existing tree cover and seeks to promote forestry and agroforestry as appropriate land uses for scrub lands and areas in need of conservation.

In pursuing this policy, the aim will be to allocate an adequate area of suitable land to forestry and agroforestry to ensure conservation of soil and water resources, sustainable income and employment for rural people, and integrated crops and livestock production. This will require a programme of land use survey, classification and planning to effect proper subdivision and lease-hold settlements as incentives towards encouraging greater tree coverage on farms.

Forest and Watershed Rehabilitation

The objective of this project is to establish tree cover on degraded watersheds and sand-mined lands in order to reduce erosion, improve water retention and enhance rural development through agroforestry practices and reforestation.

Forest based industrial development-the proposed policy seeks to ensure production of future supplies of wood raw material and utilization of existing mature timber for industrial purposes, while maintaining the amenity value of trees in urban areas.

Utilization of Neem

The objective of this project is to study the prospects for utilizing various products of the neem tree with a view to industrial development.

Fuelwood and energy-the proposed policy seeks t-explore opportunities to utilize fuelwood for energy purposes and to promote feasible wood energy applications in appropriate industries.

The aim would be to develop a sustained demand for wood energy by domestic and industrial users, whether as fuelwood or as charcoal, both to promote the use of agroforestry products and to conserve other energy sources.

The objective of this project is to determine the production and demand for charcoal and to evaluate the opportunity to expand the market.

Conservation of forest ecosystems-this policy seeks to maintain biological diversity and to protect forest ecosystems and natural habitats, while sustaining the flow of social, economic and environmental benefits derived from forests for nationals and tourists.

The aim is to increase public awareness of environmental problems and the need to conserve the natural resources of the island as part of the nation's heritage. In the absence of a coordinated system of management of existing habitats, the following project is extremely important:

3

FOREST BIODIVERSITY VALUES AND FUNCTIONS

INTRODUCTION

There is a vast body of knowledge on forest biodiversity among commu-nities in Orissa, i.e. knowledge of occurrence and distribution of plant and animal species and functions and uses of these (Singh et al., 1999). The communities visited in the PMDFO preparatory phase empha-sised the importance of ecosystem functions for forest health and for the health and fertility of surrounding areas (forests were seen as important for e.g. regulation of hydrological cycles and for provision of humus and nutrients for surrounding fields) (ibid.).

Both "use-values" and "non-use-values" of biodiversity were impor-tant in the communities visited. The communities were proud of their protected forest areas and cared for the forest not only for its production of goods but also (and explicitly in higher caste groups) for its existence value. Parts of some of the protected forests areas were sacred groves and carried important religious functions. Tribal people in the villages invari-ably said that life without the forest would be unthinkable. Their identity is linked to the forest.

A reference may be made here to all the different life forms in the forest, the animals, plants, fungi and microorganisms. Just like the humans, the individual lives in the forest also have a stake in its contin-ued existence for their own being and regeneration. They may be re-garded as stakeholders in their own right.

In most communities in the forested parts of Orissa, forest products provide an important part of subsistence and income, particularly for the tribal groups. Trees provide wood for construction of houses, tools and fuelwwod. Non-timber forest products (NTFPs) from a large diversity of species are collected and used for domestic consumption or sold. Until 2000, the rights to trade and process NTFPs were leased in a monopoly contracting system to a small number of traders and compa-nies, some of them state owned. Local communities had no such rights and people were forced to sell their NTFPs to defined outlets, often for a pittance (Hobley and Shields, 2000). This situation has now changed, partly as a result of the PMDFO process. Both NTFPs sold and NTFPs used domestically are vital for the liveli-hoods of Orissa's forest dwelling communities (Mallik et al.,1998).

For example, green leaf vegetables, small game, fruits and tubers provide proteins, minerals and vitamins which would otherwise be insufficient in the local diets, and they are of key importance during food shortages. Mushrooms form an important part of the diet during the 2-4 months when they are available, particularly for tribal groups. Particularly tribal people may spend a few hours daily for mushroom collection during this period. In many of the villages, mushrooms are sold at the local market or to other households in the village (while tribal women may spend hours daily in the forest, non-tribal women in the same village may seldom or never enter the forest).

PROJECT DESCRIPTION

In 1997, Orissa Forest Department embarked on preparation of the project titled "Capacity Building for

Participatory Management of Degraded Forests in Orissa" (in this report referred to as PMDFO) with funding from Sida. Scandiaconsult Natura AB and Asian Forest Network assisted Orissa Forest Department in the project preparation phase, which took place during the period January 1998-May 1999.

The PMDFO project aims at promoting sustainable and community based management and use of forests in Orissa and thus contribute to social and economic development in Orissa.

The preparatory phase of the project (Dec 1997-May 1999) aimed at developing background and capacity for a longer support in the second phase of implementation. The goals of the preparatory phase as expressed in OFDs final inception report (Scandiaconsult Natura AB/Asia Forest Network, 1998a) were:

1. Further elaboration of JFM and CFM concepts by exploring relation-ships between joint and community based forest management prac-tices;
2. Strengthening the capacity, including restructuring, of the Orissa Forest Department;
3. Knowledge about low cost methods for reforestation through natural regeneration;
4. Use of the information and learning experiences generated during this period for the preparation of the project document for phase II.

The expected results of the first project phase were:

(a) a project document and work plan for the next 3-5 year phase;

(b) reports from a number of studies related to the project goals;

(c) an established and operative policy cell within OFD;

(d) a plan for the reorganisation of OFD;

(e) a plan describing how to study the succession of natural regeneration and its management, and how to incorporate lessons learnt within OFD;

(f) training carried out for staff and village representatives.

The following studies were carried out in the preparatory phase of the PMDFP project:

- Qualitative assessment of community initiative in forest management
- Benefit sharing mechanism between OFD and communities
- Legal framework and institutional arrangement between OFD and communities
- Identification of research areas and preparation of prioritised research agenda and research approach based on participatory concept
- Study on actors of relevance for implementation of themes of the project
- Study on entry points of relevance for implementation of themes of the project
- Silviculture and management practices in community based forest management initiatives
- Study on biodiversity aspects
- Study on domestic and commercial use of NTFPs, excluding marketing
- Study on domestic and commercial use of NTFPs, including marketing
- Identification and testing of criteria for area selection in forests not protected by communities
- Monitoring and evaluation for CFM and JFM areas

Reports from the studies have been presented, and a base paper has summareised the resommendations from the studies (Scandiaconsult Natura AB/Asia Forest Network, 1998b).

A project document for the implementation phase has been presented to Sida for consideration of support (OFD, 1999). Due to delays of OFD reorganisation for various reasons during the preparatory phase, a policy cell within OFD has not been established but is proposed in the project document. The same applies to training of staff and village representatives. The implementation phase of the project "Capacity Building for Participatory Management of Degraded Forests in Orissa" has not yet started, and funding is still sought. A complicating factor here is the withdrawal of funds for Swedish development cooperation with India after the Indian nuclear bomb tests in 1998.

ANALYSIS OF LESSONS LEARNT IN RELATION TO BIODIVERSITY ASPECTS OF PMDFO

As in any complex project development process with many actors in-volved, the lessons that can be drawn are innumerable. For those inter-ested in the results and recommendations from the different studies, the study reports and the base paper summarising the results provide useful reading.

The intention here is not to analyse the results of the preparatory phase of the project in relation to the objectives of the same. Instead, I will try to present some lessons which relate to the biodiversity aspects of the project. My understanding of the process and of lessons learnt are based largely on my own involvement in the biodiversity study of the preparatory project phase (Singh et al., 1999). This means inter alia that as an actor in the first project phase I am biased, and I carry my biases with me into this case study report.

In order to attempt a meaningful analysis, the lessons presented in this case study may be separated into three levels:

- lessons related to development of the project concept;
- lessons learnt in preparation of the project, i. e. major results and recommendations from the studies in the project preparation phase; and
- lessons learnt regarding the project development process.

In the following, I will try to summarise some lessons learnt relating to biodiversity aspects on these three levels.

Lessons Related to Development of the Project Concept

Sida supported a Social Forestry Programme in Orissa between 1983 and 1996. This programme covered all of Orissa and focussed on plantation of village forests on land NOT classified as forests, i e land outside the normal domain of OFD. The major costs of the programme were staff (in a separate Social Forestry wing of OFD created for the programme) and labour for tree planting in village woodlots, weeding and guarding plantations etc. Through creating seasonal employment opportunities, the programme did contribute towards poverty alleviation.

It was also relatively successful in establishing the plantations. However, the social objective of the programme, to create a sustainable income for the poorer sections of society through tree growing, was never realised to any larger extent (Ingevall, 1997; Holm & Nilsson, 1997). The Village Forest Committees which were created did not realise their actual ownership of the plantations, and distribution of benefits from these were not equitable.

The tree plantations in the Social Forestry Programme were almost exclusively monocultures of exotic tree species and hence with low natural biodiversity and production of few non-timber forest products beside fuel in the form of wood and dry leaves.

Based on the various difficulties and shortcomings of the Social Forestry Programme, Sida's support came to an end in 1996. Sida's negotiations with Orissa regarding a new forestry project has been a long process, and Sida has played an active role in supporting change within OFD.

This has been the case also in the re-orientation from plantation forestry based on monocultures in the social forestry project to the focus on natural regeneration of forests in the PMDFO project. Conservation and sustainable use of biodiversity are integral to this focus, and so is a focus on equitable sharing of benefits from the use of biodiversity. All three major objectives of the Convention on Biologi-cal Diversity have thus been actively addressed by Sida in the dialogue with OFD (Ingevall, 1997).

Ultimately, Sida's support to the project is aimed at poverty allevia-tion, which is one of the main goals of Swedish development coopera-tion. In this project, the biodiversity objectives are intimately linked to the overall aim of poverty alleviation.

Sida's active role in supporting a focus within OFD on participatory management of naturally regenerating forests has been a key factor in the development of the project concept. Sida's active role was possible because of the professional skills of Sida staff in relation to sustainable and participatory natural resource management, a long history of cooperation with OFD and other actors in Orissa, presence in India, knowledge of national and state level policies and legislation of relevance in the project context, and persistence in the dialogue with OFD. The key role of the donor in the dialogue with the cooperation partner is an important lesson in this case study.

Major Lessons from Studies in the Project Preparation Phase

The preparation of the PMDFO project was organised as a series of studies undertaken on issues of relevance for the project. A summary of lessons learnt from these studies, as presented in the project base paper. The studies did not cover issues related to reorganisation of OFD. The following is a brief summary of the major lessons which relate to the biodiversity aspects of the project.

Roles of Communities in Forest Management

In accordance with the national and state level forest policies, the broad objectives for management of CFM/JFM areas in Orissa should be: rehabilitation of degraded forests through natural regen-eration, improving productivity particularly of non-timber forest products (NTFPs) through appropriate measures including biodiversity conservation, sustainable management and sustainable use to supply forest products to meet village communities' requirements. Within this broad framework, area-specific management objectives will be developed by each village level organisation (VLO).

The management objectives need to include an ecosystem perspec-tive, where issues related to e.g.fodder production and improvement of pastures both within and outside the protected forests may need to be addressed.

Participatory management systems need to be developed. Participa-tory forest management will include a baseline survey, a planning process using PRA methodology, and participatory monitoring and evaluation. Planning and monitoring will address managerial, finan-cial, slvicultural, social and biodiversity issues.

Policy and Legal Framework

The regulations and policies concerning NTFPs do not sufficiently recognise the rights of forest dwelling communities with respect to collection, processing and marketing.

Orissa government needs to carry out an internal review of policy and laws relating to NTFPs in order to provide clear definition of NTFPs, ensure transparent management operations in relation to collection, processing and marketing of NTFPs, sharing of revenue generated, institutional arrangements etc. The review also needs to ensure compliance with National Forest Policy and national conserva-tion guidelines.

The existing practice of monopoly leases in NTFP trade and process-ing should be discontinued. Leases could be given to a number of buyers including co-operatives, non-profit making societies, joint sector companies, VLOs and their federations. Formation of primary collectors' institutions could be helpful. The responsibility of primary collection, storage, and minor processing of NTFPs could be en-trusted to VLOs. Higher level processing of certain NTFPs could be allowed on experimental basis to capable VLOs.

The current ban on felling means locking the resource without its sustainable use. It is likely to negate the benefits of CFM/JFM in the state, and it should eventually be lifted. However, this would require the development of regulatory and monitoring mechanisms, to ensure (1) designationa and continued protection of ecologically important areas; (2) ecologically sustainable logging practices and (3) a system for fair sharing of benefits, where the private sector does not dominate the scene with minor benefits to the local communitites.

Needs for Training and Improved Knowledge

Increased production and collection of NTFPs will depend on partici-patory forest management systems where the forests are managed for production of both timber and NTFPs. Management systems that enable sustainable production and collection of NTFPs is generally expected to contribute to restoration of forests rich in biodiversity, which will be ecologically sustainable and hence contribute to sustain-able livelihoods of forest-dependent communities. However, produc-tion of some major NTFPs (e.g. kendu leaf, sal leaves and sal seed) may lead to simplified and eventually degraded ecosystems. A better understanding is needed of how best to optimise production of these NTFPs.

Appropriate training for both VLO members and OFD staff is important. Topics for training concerning silviculture and manage-ment of CFM/JFM areas may include i) importance of biodiversity in forest management, ii) natural regeneration survey and simple inventory of forest resource (number of trees and basal area per hectare), iii) sustainable silviculture and management practices for CFM/JFM areas, iv) management plan preparation, v) sustainable harvesting, use and marketing of products from CFM/JFM areas, and vi) community organisation and functioning of VLOs.- Preparation of a field manual on silviculture and management of CFM/JFM areas is essential for training as well as for extension support. Appropriate literature on silviculture and management of CFM/JFM areas needs to be published in Oriya for use by VLOs, OFD staff and NGOs.

The Biodiversity Study

The biodiversity study of the PMDFO project preparatory phase con-sulted members of 27 communities in 4 districts in Orissa. The bio-diversity aspects studied included:

- biodiversity and ecosystem functions in forests protected by local communities;
- collection and use of non-timber forest products (NTFPs) by different groups in the communities;
- study of social, cultural and spiritual values of biodiversity; and

- monitoring of biodiversity as an integral part of participatory forest management.

A summary of findings on the first three aspects has been presented earlier in this report. Below are some of the conclusions concerning biodiversity monitoring in participatory forest management. Biodiversity monitoring in participatory forest management Participatory forest management with integration of biodiversity aspects and for optimal yield of NTFPs would typically include management choices like selection of both timber and multi-purpose tree species in silvicultural management; avoidance of indiscriminate clearing of forest under-story, and keeping NTFP species in cases where under-story clearing is necessary (Scandiaconsult Natura AB/Asia Forest Network, 1998c).

Some NTFP collection, like sweeping of leaves from the forest floor (for fuelwood or to facilitate collection or e.g. sal seeds), harvesting of tubers and medicinal plants, and hunting, may well have negative effects on forest biodiversity and ecosystems if done indiscriminately. There is for instance evidence that sweeping of leaves has negative impact on development of forest soils in some protected forests.

In order to learn about the effects on biodiversity of management choices made, monitoring is needed. The best approach to biodiversity monitoring in participatory forest management, based on consultations with 27 communities, was found to be simple inventories done as transect walks through protected forest areas, documenting plant species and arthropod groups, complemented by an enumeration by the community of plant and animal species known to live in the forests, and trends (increasing or decreasing in numbers) for key species.

The biodiversity-NTFP study recommended that this simple biodiversity monitoring be integrated with the communities' participa-tory forest management planning as recommended by the study on participatory monitoring and evaluation of forest management (Balasubramanian, 1998). Biodiversity monitoring should thus be done by the community for its own needs. Detailed design of biodiversity monitoring can hence only be done by the communities themselves. This should form part of the implementation phase of PMDFO.

Lessons Learnt Related to the Project Development Process

Most of the studies carried out in the preparatory phase of PMDFO were done in consultation with local communities. The participatory approach was planned for at the inception of the preparatory phase. Consultations between local communities, NGOs and OFD in the form of seminars etc. were also planned in order to discuss the outcome of the different studies and constitute a basis for further project planning. Consultations were held between OFD and the major NGOs active in supporting CFM/JFM in Orissa in the form of a number of seminars during the preparatory phase of the project. This provided useful oppor-tunities for exchange of views and sharing lessons regarding CFM/JFM between OFD and the NGOs.

Efforts were made to integrate the work of the different studies. For instance, the consultants carrying out the biodiversity-NTFP study worked actively to integrate this with the other studies. This was done by consultations with the consultants responsible for the other studies and by efforts to integrate preliminary results of other studies into the biodiversity-NTFP study and vice versa. Some of the field work under-taken in the biodiversity-NTFP study was done jointly with the study of silviculture and forest management. This was beneficial for mutual learning about relationships between biodiversity, silviculture and forest management.

However, consultations which include local communities and their federations regarding the outcome of the studies have not as yet been carried out. Without such consultations, the results and recommenda-tions from the preparatory phase can only be seen as preliminary. One of the factors restricting the level of consultation in the review stage of project preparation was shortage of time in the project planning process. This particularly restricted the possibilities to provide informa-tion to communities in the form of study outcomes and proposals as well as the opportunity of communities to comment on these. It was origi-nally planned to translate the study reports and proposals into Oriya. This, however, was not done due to shortage of time.

A major lesson learnt here is that a process for participatory review should be planned at the outset which guarantees that the different interest groups have opportunity and time to present their views on the outcome, and that their views are taken into account. To enable this, time must be set aside for participatory review at the end of the process in the original time plan (and any delays should not be allowed to en-croach on the time allowed for participatory review); a clear process for participatory review must be agreed on and adhered to; and the neces-sary documents need to be translated into the local language and distrib-uted in a timely fashion to the relevant interest groups. Since a participa-tory review was not held during the preparatory phase of the project, it could form part of the preparation for a possible continuation of the project.

Diffcrent views Revealed

In the studies, one interesting observation from the joint field work of the biodiversity-NTFP team and the silviculture/ forest management team was that during the field visits, the dialogue of the two teams with community members yielded very different views on forest management. For example, when walking through a forest area protected by the community and discussing with community members, the silviculture/forest management team got the clear impression that sal trees would be favoured in a thinning operation, while wild fruit tree species would be felled. Conversely, the biodiversity team were told that sal trees and wild fruit trees were equally important in forest management.

This points towards the need for great care in interpretations of results from brief visits to local communities, and towards the need for longer term interaction, where both parties get to know each other and find space for open dialogue. It also points towards the need for dialogue with members from different groups of the same community, since needs and interests necessarily differ between these

One reason for the different views disclosed in the two groups is probably the desire of the community to please the visiting team and fulfil the team members' expectations, in accordance with the focus and emphasis they provide in the brief interaction with the community.

In spite of these shortcomings, the NGO and community level consulta-tions in the PMDFO preparatory phase helped to facilitate the formation of the state-level forum of village level communities in Orissa, which was formed in March 1999. The separate support from Sida to NGOs supporting participatory forest management in Orissa also facilitated this process.

The main objective of the Odisha Jungle Mancha, or Orissa Forest Forum (OFF), is to act as a pressure group for formulation of a pro-people forest policy favourin, the interest of forest protecting and man-aging village communities (Pal, 2000). OFF is now recognised by the Orissa Government as an advisor and acts as a bridge between the communities and the State. OFF's lobbying has enabled the formulation of Orissa's new NTFP Policy. Announced during 2000, the NTFP Policy gives rights over NTFPs to Panchayats, and what was previously a monopoly trade is now decentralised, with a minimum price back-up from the State. The communities can now freely trade and undertake value addition of NTFPs.

Although this development was only indirectly a result of the PMDFO preparatory phase, it is nevertheless a major achievement which the PMDFO process helped facilitate.

Environmental Impact Assessment of the Project Document

As agreed in the planning of the preparatory phase of the project, and in accordance with Sida's EIA Guidelines (Sida, 1998a), a brief EIA was done for OFD of the project document of May, 1999. The EIA was done by one of the consultants in the biodiversity study (Marie Byström) as a desk study of the project document. The EIA thus built on the knowl-edge gained in the biodiversity study and on previous consultations with local communities and consultants carrying out the other studies. In one sense, the environmental aspects were integrated in the planning process, and the EIA served the purpose of checking that the project document incorporated all relevant lessons learned in the planning process.

There was however no feedback process to ensure that OFD reviewed the EIA. In future planning processes, this should be ensured.

The Convention on Biological Diversity (CBD)

The three overall objectives of the Convention on Biological Diversity (CBD) are:

- the conservation of biological diversity
- the sustainable use of its components

- the fair and equitable sharing of the benefits arising out of the utilization of genetic resources.

The parties to the CBD are requested to act towards the fulfilment of these objectives at the national and local level and through international cooperation. Both the Indian national forestry legislation and the Orissa state level forestry legislation referred to above address all three major objectives of the CBD to some extent. The new NTFP Policy in Orissa (Gov of Orissa, 2000) will substantially increase the rights of local communities to process and market NTFPs, and local communities will hence hopefully share a larger proportion of the benefits from the use of NTFPs, which may at least partly be defined as genetic resources.

An Indian national biological diversity law, which more directly addresses the objectives of the CBD, is in the process of legislation. Among the features of the bill are the proposal of People's Biodiversity Registers for the documentation of local biodiversity knowledge, and of a National Biodiversity, State Biodiversity Boards and local Biodiversity Management Committees to regulate access to biodiversity and sharing of benefits.

The bill has been developed in a participatory process. Improvements have been suggested by e.g. Madhav Gadgil et al. (Gadgil et al. 2000) to further focus the biodiversity law on conservation, sustain-able use and benefit sharing of biodiversity, and to strengthen the provi-sions for local management. The biodiversity bill, when passed, may give new opportunities to local communities in terms of sharing the benefits of commercial uses of biodiversity provided through their knowledge. In addition, the People's Biodiversity Registers may assist local communi-ties to strengthen their knowledge on local biodiversity and may also help communities develop management plans, where sustainable use of biodiversity is an integral part.

The PMDFO project can be seen as one of India's activities towards the fulfilment of the objectives of the CBD. The planned project specifi-cally addresses the objectives/ obligations of e.g. CBD articles 5 (interna-tional cooperation), 6 (b) (integration of conservation and sustainble use of biodiversity into relevant sectoral or cross-sectoral plans, programmes and policies), 8 (f), (i), (j) (rehabilitation and restoration of degraded ecosystems, provision of conditions to enable conservation and sustain-able use of biodiversity, and maintenance of knowledge, innovations and practices of local communities and equitable sharing of benefits from these), and article 10 (b), (c), and (d) (adoption of measures relating to the use of biological resources to avoid or minimise adverse impacts on biological diversity, protection and encouragement of customary use of biological resources in accordance with traditional cultural practices that are compatible with conservation or sustainable use, and support to local populations to develop and implement remedial action in degraded areas where biological diversity has been reduced).

On Sida's side, cooperation in the development of PMDFO supports fulfilment of Sweden's obligations in e.g. CBD articles 5 (international cooperation), 18 (technical and scientific cooperation, in particular 18.2.: cooperation in development and implementation of national policies, human resources development and institution building, and 18.4. encour-agement and development of methods of cooperation for the use of technologies, including traditional technologies, in pursuance of the objectives of the CBD), and article 20.3. (provision of financial resources related to the implementation of the CBD through bilateral channels).

TRIPs

WTO's agreement on trade-related intellectual property rights (TRIPS) sets minimum requirements for the patentability of life forms and hence relates to access and benefit-sharing of biodiverstiy. Article 27.3 (b) of TRIPs reads: Members may also exclude from patentability; (b) plants and animals other than microorganisms, and essentially biological processes for the production of plants and animals other than non-biological and micro-biological processes. However, Members shall provide for the protection of plant varieties either by patents or by an effective sui generis system or by any combination thereof. The provisions of this sub-paragraph shall be reviewed four years after the entry into force of the WTO agreement.

Although it is very difficult to quantify the effects of this request in TRIPs for intellectual property rights on certain life forms, it is beyond doubt that the extension of intellectual property rights to life forms have direct consequences for the distribution of benefits from genetic resource use. In particular, they contribute to the present structural change toward privatisation and concentration of biological research. While this may not directly affect local communities in Orissa, the TRIPs provisions may have a long term impact on their shares of benefits from use of genetic resources.

The extension of IPRs to life in industrialised countries, which pre-ceded TRIPs, has largely taken place without a political level discussion of the proper balance between the need to stimulate technological development and other societal objectives. The balance created by present patent practice in developed countries, and codified in the TRIPs agreement, is now broadly questioned by governments and NGOs in developing countries, including India, from an ethical and socio-eco-nomic perspective. It would be advisable for developed countries to acknowledge this fact and return to a renewed consideration of their standpoints from this insufficiently explored angle. India is among the countries that have pointed to a potential conflict between their commitments under the CBD and under TRIPs, as the CBD stipulates that IPRs must be supportive of the Convention objec-tives, while TRIPs lacks reference to either conservation, sustainable use, or fair and equitable benefit sharing.

Among the concrete proposals given to WTO from developing coun-tries are the withdraw of the requirement of IPRs on life forms alto-gether, direct inclusion of some of the CBD objectives in the TRIPs text, and amending the text with explicit provisions for community rights protection. This would strengthen India's power to implement the benefit-sharing provisions in the proposed biodiversity law in an interna-tional perspective. Ultimately, this could benefit local communities in their forest management in Orissa and would hence support the objectives of the project PMDFO.

RELATIONS TO EC BDP'S GUIDING PRINCIPLES'

As part of the "Strategic Framework" developed for the EC/DFID/ IUCN Biodiversity in Development Project, seven guiding principles have been developed (BDP, 2000).

BDP Guiding principles to ensure that all development cooperation actions are sustainable and effective, and give biodiversity proper consideration

1. Adopt an ecosystem perspective and multi-sectoral approach to development pro-grammes (taking account of impacts on adjacent and down-stream areas).
2. Ensure/encourage full stakeholder participation, including partnerships with civil society, government and private sector.
3. Ensure that development cooperation projects and programmes are consistent with the wider policy framework, and/or changes are made for supportive policies and laws.

4. Ensure that institutional arrangements are effective, transparent, accountable, inclusive and responsive.
5. Promote fair and equitable sharing of costs and benefits from biodiversity conservation and sustainable use, at local and national and international levels.
6. Provide and use accurate, multi-disciplinary information, which is both accessible to and understood by all stakeholders.
7. Development cooperation investments must be sensitive to, and complement local/ national structures, processes and capacities.

Below, brief comments are given on (A) findings in the case study as they relate to the Guiding Principles, and (B) the relevance of the BDP Guid-ing Principles to mainstreaming of biodiversity in development coopera-tion as judged from this case study. (The Guiding Principles were not developed at the time of preparation of the PMDFO project.)

A. Brief comments on findings in the case study as they relate to the Best Practice Principles:

1. Adopt an ecosystem perspective and multi-sectoral approach to development programmes (taking account of impacts on adjacent and down-stream areas).

   Comment: An ecosystem perspective was part of the project approach. The notion of positive impacts on adjacent and down-stream areas was an integral part of the motivation of many communities to protect forest areas, according to the biodiversity study.

2. Ensure/encourage full stakeholder participation, including partner-ships with civil society, government and private sector. Comment: Full stakeholder participation was aimed for in the project. The work

in most of the different studies was done in consultation with local communities, and OFD had consultations with several NGOs during the preparatory phase. However, although working with participatory processes, the first phase had some shortcomings in relation to this principle.

This was the case in particular in the review of results from the preparatory phase. To what extent has the private sector been involved in the studies? This sector and its role in sustainable forestry management are not much mentioned in the report.

3. Ensure that development cooperation projects and programmes are consistent with the wider policy framework, and/or changes are made for supportive policies and laws.

   Comment: This was one of the guiding principles in the development of the project concept. The principle was also addressed by several of the studies, in the call for review of policies and laws related to NTFP production and marketing.

4. Ensure that institutional arrangements are effective, transparent, accountable, inclusive and responsive.

   Comment: The institutional arrangement of the project itself, with the project coordinator and Team Leader from the consultant, Ulf Öhman, stationed at OFD headquarters in Bhubaneswar and work-ing very closely with a group of officers at OFD, met all these princi-ples to an unusually high degree. This is likely to in itself stimulate OFDs work at reorganisation of the Forest Department in line with the above principles.

5. Promote fair and equitable sharing of costs and benefits from biodiversity conservation and

sustainable use, at local and national and international levels.

Comment: The project "Capacity Building for Participatory Manage-ment of Degraded Forests in Orissa, India" aims at promoting sus-tainable and community based management and use of forests in Orissa. This overall aim includes the promotion of fair and equitable sharing of benefits from biodiversity conservation and sustainable use at state level in Orissa. This principle was also clearly addressed by nearly all of the studies in the preparatory phase of the project.

6. Provide and use accurate, multi-disciplinary information, which is both accessible to and understood by all stakeholders.

   Comment: Although working with participatory processes, the first phase had some shortcomings in relation to this principle, as discussed in section 6.3. This was the case in particular in the review of results from the preparatory phase. This study recommends for future work in the project (1) translation of relevant documents into the local language, and (2) timely distribution of information to allow review by all relevant stakeholders.

7. Development cooperation investments must be sensitive to, and complement local/national structures, processes and capacities.

   Comment: This was a very central guiding principle in the development of the project concept. In contrast, the earlier Sida-supported Social Forestry Project in Orissa did not have sufficient sensitivity to local and national structures, processes and capacities. This was one reason for Sida to end the support to this project.

B. Comment on the relevance of the BDP Best Practice Principles to mainstreaming of biodiversity in development cooperation:

The best practice principles are excellent guiding principles for develop-ment cooperation. The fifth principle, "Promote fair and equitable sharing of costs and benefits from biodiversity conservation and sustain-able use, at local and national and international levels" is directly related to mainstreaming of biodiversity in development projects. All the other principles should be guiding principles for development cooperation in general, and the fifth principle should be promoted not only in "biodiversity projects" but also in work towards the mainstreaming of biodiversity aspects more widely in development cooperation in the natural resources management sector.

PROJECT/PROGRAMME DESCRIPTION

Brief description of the present set-up of the programme:

(a) description of the aims and activities of the project

(b) geographical area

(c) socio-economic and cultural context

(d) stakeholder analysis including ethnic and gender relations, and also an assessment of stakeholders that were "excluded" from the project activities

(e) institutional set-up including policy framework on biodiversity

(f) programme activities.

Problem Analysis

Analysis of biodiversity aspects within the programme, both historically (since inception of programme) and presently. The analysis should:

(a) be made using the following documents as analytical tools:

- "Biological Diversity-guidelines for Sida support for the sustainable use and conservation of biodiversity", Sida, 1994
- "Sida and the Convention on Biological Diversity", Sida, 1998
- "Guidelines for Environmental Impact Analyses in development cooperation", Sida, 1998
- Sida's four action programmes
- "Strategic Framework" developed for the Biodiversity in Development Project (to be able to influence the BDP-process).

(b) explicitly consider and comment on the full range of biodiversity issues in the project area, (e.g. evidence of irreversible losses, opportu-nities for improved use of biodiversity for human development, etc), to what extent biodiversity has been "mainstreamed", which biodiversity aspects that have been considered, and the relevance of these to the different stakeholders, how the aspects of biodiversity has been handled and from which stakeholders,

(c) identify key constraints to improved biodiversity management and improved human development, and assumptions made about the links between the two, within the context of the project aims and activities

(d) include assessment of whether project/programme successes have been, or are, sustainable (considering mainly aspects in connection to biodiversity use) in the long term (i.e. without further project input),

(e) explicitly consider and comment on methods used for mainstreaming aspects of biodiversity into the project/ programme, past and present monitoring of biodiversity aspects and tools to follow-up these aspects (environmental/biodiversity indicators).

Particular attention should be paid to the analysis of the projects impact on ecosystem services related to biodiversity, as well as the linkages between biodiversity and sustainable livelihood, culture and social relations (including ethnic, gender and age relations) and political and socio-economic structures.

Also gene policy questions (bioprospecting and eroison of domestic species, plants and animals, genetic diversity) should be analysed.

Institutional Resources

Identification of and consultation with possible other institutions of importance, shall be done, that could be a resource for the programme (regarding biodiversity).

Lessons Learned for the Programme

The lessons learned from the mainstreaming of biodiversity for the project/programme planning and implementation, including field methods, shall be summarised and discussed.

Lessons Learned for the Sida Mainstreaming Process

The lessons learned from the mainstreaming of biodiversity for project/ programme planning and implementation, including field methods, shall be summarised and discussed.

Conclusions and Recommendations for the Programme

Conclusions and recommendations for the programme:

(a) on methods for biodiversity mainstreaming within the programme

(b) on follow-up mechanisms (monitoring including possible indicators).

Conclusions and Recommendations to SIDA

Conclusions and recommendations to Sida regarding:

(a) improvements that could be made in the project/ programme plan-ning process and implementation of projects/programmes

(b) improvement of the analytical tools and guidelines, see c) above, for mainstreaming biodiversity aspects within Sida-DNRE's programmes.

(c) methods for how aspects of biodiversity can be mainstreamed into projects/programmes.

(d) how to monitor aspects of biodiversity in the programmes/projects (through environmental indicators etc).

Presentation of Case Studies in Seminar/workshop

The case studies will be presented in a seminar/workshop at the Sida office in Stockholm in January 2000, and when possible in the country where the case study has been taken place. The case studies will also, if possible, be presented at the BDP regional workshops on case studies: Cameroon (28/ 6-2/7/99); Sri Lanka (24/7-30/7/99); Botswana (6-11/9/99); Peru (27/9-1/10/99).

METHODS

The mainstreaming of aspects of biological diversity is of course de-pendent on the interest of the "owner" of the project/programme in extending the focus of the project/ programme. The mainstreaming should be initiated in close connection with regular reviews of Sida's support.

The consultants performing the case studies will be responsible for developing specific methods for each case study, as a contribution to lessons learned, in cooperation with the recipient partner. The consult-ants shall try to coordinate the activities in the case studies with the BDP, see also Terms of Reference for BDP case studies, appendix 2. The work on the case studies can be done in the following way:

(1) Discussion and planning in consultation with programme officers concerned.

(2) Planning of each study in consultation with the partner in coopera-tion, institutions of importance for the issue in the recipient country and consultants responsible for the implementation of each pro-gramme.

(3) Implementation in consultation with, and preferably together with, the partner in cooperation and institutions of importance for the issue in the recipient country, if possible in the field.

(4) Review of results in seminar form (and in a written report) in connec-tion with annual reviews or suchlike, and also with the staff at Sida-DNRE. Discussion of results and recommendations for each pro-gramme.

(5) The follow-up of the first mainstreaming studies should take place after approximately one year, in connection with the normal follow-up of the programme. At this point in time the methods can be evaluated and further inputs to the analytical tools and guidelines can be made. From our own results and from the BDP results, further recommendations can be made on how the work of mainstreaming aspects of biodiversity can be continued in other projects/pro-grammes in the natural resources sector.

4

FOREST PRODUCTIVITY

INTRODUCTION

No uniform set of silvicultural and management practices should be followed for all areas. Silviculture and management practices should suit the silvicultural requirements of the crop and should be adjusted to meet the objectives of management. The broad objectives for management of CFM/JFM areas should be: rehabilitation of degraded forests through natural regeneration, improving productivity particularly of NTFPs through appropriate measures including biodiversity conservation, sustainable management and sustainable use to supplement supplies of forest products to meet village communities' requirements. Within this broad framework, area-specific management objectives will be developed by each VLO.

Natural regeneration will be relied on to regenerate forests. No gap planting is recommended. The gaps and blank areas in CFM/JFM areas should be developed for grass production by cutting tall weeds and bushes not yielding any fodder. Fodder is an important requirement of village communities and development of such areas for grass production will be better than planting of trees. An efficient participatory learning and monitoring approach needs to be adopted not only to develop appropriate management practices, but also to ensure their proper implementation.

A baseline survey through which the status of the forest and socio-economic conditions of the village will be analysed through PRA before the preparation of the management plan.Participatory monitoring and evaluation (PAME) should be done to review the progress of the forest management. It may be undertaken once in a year with participation of both VLO and OFD field staff. PAME would address managerial, financial, slvicultural, social and biodiversity issues. Strong extension support to be provided by OFD ad NGOs is needed for proper management of CFM/JFM areas. Special training will be necessary for OFD field staff (see further below). The current ban on felling (in other words locking the resource without its sustainable use) is likely to negate the benefits of CFM/JFM in the state. Its adverse effects include:

- lower productivity of the forest resource;
- increased incidence of illicit felling to meet villagers' requirements;
- reduced quality of growing stock as a result of illicit cutting of better stems and leaving the inferior ones as was observed during field visits, and
- mistrust between villagers and OFD because of the feeling gaining ground among the villagers that FD is interested in getting the JFM forests protected and rehabilitated and not in giving them the prom-ised forest products; such mistrust can be suicidal for JFM pro-gramme.

Sustainable use of NTFPs

The collection and use of NTPF provide an important part of subsist-ence and income, particularly for the tribal groups. NTFPs regularly collected for domestic consumption include fuelwood, mushrooms, medicinal plants, green leaf vegetables, fruits, tubers, resin, fibres, thatch-ing grass, and dyes. Important NTFPs collected and sold commercially include kendu leaves, siali leaves, sal leaves, sal seed, and mahua flowers. Present levels and methods of collection of NTFPs seem to be overall sustainable, with considerable potential for increased production and collection.

Increased production and collection of NTFPs will depend on partici-patory forest management systems where the forests are managed for production of both timber and NTFPs. Management systems that enable sustainable production and collection of NTFPs generally is expected to contribute to restoration of forests rich in biodiversity. However, production of some major NTFPs (e.g. kendu leaf, sal leaves and sal seed) may lead to simplified and eventually degraded ecosystems, if production is not carefully managed on a sustainable basis.

A better understanding of how best to optimise production of these NTFPs is, therefore, important. The regulations and policies concerning NTFPs do not sufficiently recognise the rights of the forest dwelling communities with respect to collection, marketing and processing. While there is scope for a manifold increase of the production and value addition of NTFPs, it will depend on policy and legal changes.

The issues in NTFP trade relate to ownership rights over forest land and produce, accessibility, forms of disposal, marketing arrangements as well as procurement of raw material and price fixation. The existing practice of monopoly leases in NTFP trade and process-ing should be discontinued. Once the royalty and minimum support price etc. are decided and regulated, lease could be given to a number of buyers including co-operatives, non-profit making societies, joint sector companies, VFCs, VSSs and their federations. Formation of primary collectors' institutions could be helpful. The responsibility of primary collection, storage, minor processing etc. of NTFPs could be entrusted to VLOs. Higher level processing of certain NTFPs could be allowed on experimental basis to capable VLOs. Primary collectors/VLOs may be permitted to sell a major part of their finished forest products to ORMAS.

Legal/policy Issues

Orissa government needs to carry out an internal review of policy and laws relating to NTFPs in order to provide clear definition of NTFPs, ensure transparent management operations in relation to collection, processing and marketing of NTFPs, sharing of revenue generated, institutional arrangements etc. The review also needs to ensure compliance with National Forest Policy and national conservation guidelines. Policy constraints include:

- the provisions in JFM resolutions (1993 & 1996) regarding the roles of VLOs and FD need clarity to empower the VLOs to implement the programme and the FD to assist and facilitate implementation;
- usufruct sharing mechanism requires clarity, and major harvest needs a clear definition;
- a provision appears necessary to enable sale of products in excess of village subsistence needs, and appropriate marketing system needs to be developed to improve the returns to VLO members, particularly the poor, from CFM/JFM;
- harvesting trees should be the responsibility of the VLOs; the govern-ment resolution on JFM is not clear as to whether the trees to be removed in thinning are to be harvested by VLO members or by FD;
- ban on fellings in CFM/JFM areas needs to be removed to meet the requirements of management plans;
- CFM/JFM areas need to be taken out of the purview of working plans.

Training

Appropriate training for VLO members and OFD staff is important. Topics for training concerning silviculture and management of CFM/ JFM areas may include i) importance of biodiversity in forest manage-ment, ii) natural regeneration survey and inventory of forest resource (number of trees and basal area per hectare), iii) sustainable silviculture and management practices for CFM/JFM areas, iv) management plan preparation, v) sustainable harvesting, use and marketing of products from CFM/JFM areas, and vi) community organisation and functioning of VLOs.

Preparation of a field manual on silviculture and management of CFM/JFM areas is essential for training as well as for extension support. Appropriate literature on silviculture and management of CFM/JFM areas needs to be published in Oriya for use by VLOs, OFD staff and NGOs.

The training programme to be designed for phase II needs to inte-grate biodiversity aspects in all its parts. Training needs related to biodiversity include i) training of VSS and VLOs in participatory man-agement, monitoring and evaluation of biodiversity and NTFPs, ii) training of FD at central level, divisional level and local level in participatory management, monitoring and evaluation of biodiversity and NTFPs, and iii) training and skills development at village level for value addition of NTFPs, including packaging, storing, marketing, accounting and other management skills. Appropriate training for NGOs also should be arranged.

Training should be conducted as much as possible in the field. Training in skills development at village level for value addition and marketing of NTFPs needs to be developed together with local villagers and NTFP producers. Successful cases of local value-adding and market-ing of NTFPs may be used as examples in this training.

AUTHORITY AND PURPOSE OF THE FOREST LEGACY PROGRAM (FLP)

A. Authority

The Cooperative Forestry Assistance Act (CFAA) of 1978, as amended, (16 U.S.C. 2101 et. seq.) provides authority for the U.S. Secretary of Agriculture (Secretary) to provide financial, technical, educational, and related assistance to States, communities, and private forest landowners. Section 1217 of Title XII of the Food, Agriculture, Conservation and Trade Act of 1990 (P.L. 101-624:104 stat.3359; 16 U.S.C. 2103c), also referred to as the 1990 Farm Bill, amended the CFAA and directs the Secretary to establish the FLP to protect environmentally important forest areas that are threatened by conversion to nonforest uses. This authority continues indefinitely. Through the 1996 Farm Bill (Federal Agricultural Improvement and Reform Act of 1996; P.L. 104-127; Title III-Conservation; Subtitle G-Forestry; Section 374, Optional State Grants for Forest Legacy Program), the Secretary is authorized, at the request of a participating State, to make a grant to the State to carry out the FLP in that State, including the acquisition by the State of lands and interests in lands.

B. Purpose of the Forest Legacy Program

The purpose of the FLP is to ascertain and protect environmentally important forest areas that are threatened by conversion to nonforest uses.

FLP seeks to promote forestland protection and other conservation opportunities. Such purposes shall include the protection of important scenic, cultural, fish, wildlife and recreational resources, riparian areas and other ecological values. Traditional forest uses, including timber management, as well as hunting, fishing, hiking, and similar recreational uses are consistent with purposes of the FLP. Both purchased and donated lands and interests in lands through the use of conservation easements and fee-simple purchase are used to acquire forested land meeting Forest Legacy purposes from willing sellers or donors.

C. Delegations of Authority

The Secretary has delegated authority to administer all aspects of the FLP to the Under Secretary for Natural Resources and Environment (7 CFR 2.20(a)(2)(xvi)) who in turn has delegated the authority to the Chief of the Forest Service (7 CFR 2.60(a)(16). Delegations only apply within the U.S. Department of Agriculture and its agencies. The role of State and Regional programs, and the right of States to elect the State Grant Option, are contained in the authorizing statute and these program implementation guidelines.

Description of Terms and Abbreviations

Assessment of Need (AON) is a document produced by a State, or a federally recognized Indian Tribe, in consultation with the State Forest Stewardship Coordinating Committee (SFSCC). The AON contains the an assessment of the forests and forest uses, a description of forces that are converting forests to nonforest uses, describes Eligibility Criteria developed by the State to identify important forest areas to be proposed as Forest Legacy Areas (FLA), and acts as a guide to implementation of FLP in the State.

Assessment of Need (AON) Amendment is a document produced by a State to amend their AON, to add or delete Forest Legacy Areas (FLA), or to modify the Eligibility Criteria. CFAA is the Cooperative Forestry Assistance Act of 1978, P.L. 95-313, 92 stat. 365, 16 U.S.C., 2101 et seq. (as amended through P.L. 107-195, June 16, 2002).

Conservation Easement is a legal agreement a property owner makes with a governmental entity or a nonprofit organization to restrict activities allowed on the land in order to protect specified conservation values. Easement restrictions are tailored to the particular property and to the interests of the individual landowner. All FLP conservation easements are held in perpetuity.

Eligibility Criteria are a set of factors developed by the State lead agency, in consultation with the State Forest Stewardship Coordinating Committee (SFSCC), to evaluate geographic areas to determine if they contain significant environmental values to be considered an 'important forest area' and contain "threats" of conversion to be eligible as a Forest Legacy Area (FLA).

Federal Appraisal Standards are those standards contained in the publication entitled "Uniform Appraisal Standards for Federal Land Acquisitions: Interagency Land Acquisition Conference, 2000," or subsequent amendments or updates.

Forest Legacy Area (FLA) is a geographic area with important forest and environmental values, that satisfies identified Eligibility Criteria and has been delineated, described, and mapped in a State's AON for the FLP. Acquisition of lands and interests in lands for the FLP can only occur within approved FLAs.

Forest Legacy Area (FLA) Boundary Adjustment is a minor change to an existing FLA to create a more logical or manageable boundary.

Forest Legacy Program (FLP) Project is an individual or series of land or interest in land acquisition transaction(s). The transaction(s) can be on an individual tract or multiple tracts in a distinct geographical area. A FLP project relates to a single funding event in a given fiscal year. FLP projects can have a single parcel that can be completed at one closing or more than one parcel that can be completed in a succession of closings. If a successive FLP project is proposed on a parcel or in a distinct geographic area each transaction is treated as an independent unit in the project selection process and funding is not guaranteed.

Forest Service (FS) is the United States Department of Agriculture Forest Service. Forest Service Region/Area/IITF refers to the field units of the Forest Service responsible for FLP management and oversight within the Forest Service Regions, Northeastern Area (Area) or International Institute of Tropical Forestry (IITF).

Forest Stewardship Plans, or multi-resource management plans, are prepared with the purpose of achieving long-term stewardship of forestland. Such plans identify landowner objectives and describe actions to protect and manage soil, water, range, aesthetic quality, recreation, timber, and fish and wildlife resources, and other conservation values identified on the tract. Plans are to be prepared by a professional resource manager. A Forest Stewardship Plan that meets the requirements of the Forest Stewardship Program or a multi-resource management plan is required for FLP qualification. The State Forester or equivalent, or their designee must approve the plan.

Full Fee Purchase is a land conveyance where a purchaser acquires all rights, title and interest in a property from a seller or owner. It is also known as fee simple or fee acquisition. Geographic Regions are the collection of States that makeup the National Association of State Foresters (NASF) Regions.

The three regions are: North (consisting of the States within the FS Northeastern Area), South (consisting of all the States within the FS Southern Region, and the Territories of the International Institute of Tropical Forestry), and the West (consisting of all the States within the FS Northern, Rocky Mountain, Intermountain, Southwestern, Pacific

Southwest (including the Commonwealth of the Northern Mariana Islands, Guam and American Samoa), Pacific Northwest and Alaska Regions.

Indirect costs relate to costs of the management and administration of the FLP. Indirect costs, unlike salary, which is a direct cost, are defined as costs not readily assignable to a specific legacy acquisition.

In-kind contributions are non-cash contributions, including third-party contributions. In-kind contributions must be expenses necessary to accomplish program activities, and allowable if the Federal Government were required to pay for them.

Interests in Land are a right, claim, or legal share in real property that are less than the full title. Land Trust is a nonprofit organization, as described in 501(c) of the Internal Revenue Code of 1986, that protects land by working with landowners who wish to donate or sell fee title or conservation easements to maintain conservation values associated with the land.

Market Value is the amount in cash, or in terms reasonably equivalent to cash, for which in all probability the property would be sold by a knowledgeable owner willing but not obligated to sell to a knowledgeable purchaser who desires but is not obligated to buy. (Uniform Appraisal Standards for Federal Land Acquisitions: Interagency Land Acquisition Conference, 2000, p.4.) Multi-State Entity is a government-established organization involving two or more States or Indian tribes whose jurisdiction encompasses all or portions of the land area of an FLA(s). National Association of State Foresters (NASF) is the organization representing State forestry organizations in all 50 States, the territories and the District of Columbia.

Nonfederal Cost Share refers the nonfederal cost-share required to receive FLP funding. There are three main categories of activities that meet this requirement:

(1) the value of land, or interests in land, dedicated to the FLP that is not paid for by the Federal government,

(2) nonfederal costs associated with program implementation, and

(3) other nonfederal costs associated with a grant or other agreement which meets FLP purposes. The nonfederal cost-share must be documented, and in the case of a grant, must meet the timing, terms and conditions of the grant.

Nonforest Uses

Noncompatible-nonforest uses are uses of the land inconsistent with maintaining forest cover including, but not limited to, activities that result in extensive surface disturbance such as residential subdivisions, commercial development, and mining. These uses generally should be excluded from FLP conservation easements or land purchases. FLP funds should only be used on parcels with forestland as defined in the State's AON.

Compatible-nonforest uses are nonforest uses of the land that may be compatible with forest uses as part of an undeveloped landscape, including cultivated farmland, pasture, grassland, shrubland, open water, and wetlands. These nonforest uses should be less than 25 percent of the total area. Forest Legacy funds should only be used on parcels with forestland as defined in a State's AON. Other funding sources may be used to protect nonforested areas on those parcels with less than the minimum required forest cover.

Nontrust Allotment Lands are privately owned fee simple lands owned by tribal members and if they are forested, are eligible for the FLP when they are located within an approved FLA. Trust lands and reservations are already protected through the trust relationship between the U.S. Department of the Interior and the tribe and are ineligible for the FLP.

Pass-through describes a land transaction whereby a third party, such as a land trust, acquires interests in lands with the intent to convey such interests to a unit of government. The transaction can include a full or partial donation, or sale at market value.

Payment in Lieu of Taxes (PILT) is made by tax-exempt entities, including the Federal government, to compensate local jurisdictions for tax revenues foregone as a result of ownership by a tax-exempt owner. Any FLP tract acquired in fee and held by the FS is eligible for PILT payments (entitlement land as defined at 31 U.S.C. 6901). Federal funds for PILT are not authorized for any land or interests in land held by nonfederal entities, or for conservation easements held by the United States.

Program Funds are FLP funds that are appropriated by Congress and allocated by the FS to three categories: Project funds, Administrations funds, and AON Preparation funds. Project Evaluation Criteria are developed by the States, in consultation with the State Forest Stewardship Coordinating Committees (SFSCC), to evaluate the eligible tracts submitted by interested landowners for inclusion in the FLP.

Relocation refers to the provision in the Uniform Relocation Assistance and Real Estate Property Acquisition Policies Act of 1970 (PL 91-646 or 42 U.S.C. 4601) which requires Federal agencies and programs to pay for the relocation of a person displaced by a federally funded real estate transaction.

Reserved Areas are designated areas where nonforest uses (e.g. house, barn, remote recreation camps, etc.) are or will be allowed, but are inseparable from the land holding and do not have a detrimental effect on the conservation easement values. These areas shall be defined and described in the conservation easement and may be restricted in terms of their use, or provisions made through cost and time to cure and treatment. To the extent possible these areas of non-compliance should be excluded from the FLP project.

Reserved Interest Deed is where the grantee (government) acquires all rights, titles, and interests in a property, except those rights, titles, and interests that may run with the land that are expressly reserved by a grantor (landowner). Secretary is the U.S. Secretary of Agriculture.

State refers to any of the 50 States, Puerto Rico, Guam, the United States Virgin Islands, the Commonwealth of the Mariana Islands, and American Samoa participating in the FLP. State Forest Stewardship Coordinating Committees (SFSCC) are defined, and their duties are described, in Section 19(b) of the CFAA (16 U.S.C. 2113).

They are chaired and administered by the State Foresters, or equivalent State officials, with membership composed of representatives from the following agencies, organizations, or individuals: Forest Service; Natural Resources Conservation Service; Farm Services Agency; Cooperative State Research, Education, and Extension Service; local government; consulting foresters; environmental organizations; forest products industry; forest land owners; land trusts; conservation organizations; the State fish and wildlife agency; and others determined appropriate by the Secretary.

The SFSCC makes recommendations to the State lead agency regarding the AON, AON amendments, and the determination of project priorities.

State Lead Agency is that unit of State government responsible for coordinating the establishment and implementation of the FLP in the State, as designated by the Governor or pursuant to State law. The State lead agency is usually a forestry agency, but may be another natural resource agency.

Tribal Assessments of Need-An AON is developed by a federally recognized Indian Tribe in cooperation with the State and the SFSCC. Only nontrust allotment lands are eligible for FLP. Lands or interests in lands purchased under a Tribal FLP can be through a grant to a cooperating State or through the Federal acquisition option.

Working Lands Conservation Committee is a committee of the NASF having coordination and consultation responsibilities within that organization regarding the FLP.

National Environmental Policy Act (NEPA)

NEPA applies to certain proposed actions of the Federal Government. NEPA does not apply to the independent actions of States or private property owners. It has no applicability to a private property owner's use or development of his/her property rights, nor the development of a State's FLP. It could apply to Federal agency actions undertaken on private property if the U.S. acquired a right to permit or deny certain land uses and then proposed to exercise that right, but in such an instance it would be the U.S. that would be required to satisfy NEPA requirements, not the private owner. It should be known that:

1. A Programmatic Environmental Assessment and a Finding of No Significant Impact was completed for the national FLP and signed by the Chief of the Forest Service.
2. Under the Federal acquisition option, the FS NEPA regulations (Forest Service's Environmental Policy and Procedures Handbook 1909.15-92.1, effective 9/21/92), the acquisition of an individual Forest Legacy tract and/or easement may be categorically excluded from the preparation of an Environmental Impact Statement or an Environmental Assessment unless

scoping indicates extraordinary circumstances exist.

Coordination with State Forestry Agencies

Whereas most State lead agencies are State Forestry agencies, and the CFAA establishes a broad cooperative relationship between the FS and State Foresters, the FS shall appoint a representative to coordinate with the Working Lands Conservation Committee of the NASF (or its successor) regarding the FLP. Periodically, the Director of Cooperative Forestry, and the appointed FS representative shall meet with the NASF Working Lands Conservation Committee to assess program operations, accomplishments, and policies. In States where the State Forestry agency is not the designated State lead agency for the FLP, a coordinating mechanism shall be instituted between the State lead agency, the State Forester, and the SFSCC.

Assessment of Need (AON) and Identification of Forest Legacy Areas (FLA)

A State or a federally recognized Indian tribe conducts an AON, in cooperation with the SFSCC, to document their need for inclusion in the FLP, through an evaluation of current forests, forest uses, and the trends and forces causing conversion to nonforest uses. Federally recognized Indian Tribes must cooperate with the SFSCC when conducting an AON for nontrust allotments lands.

The AON is intended to define the Eligibility Criteria to be used in the identification of important forest areas to be proposed as an FLA; identify and delineate the boundaries of forest areas meeting the Eligibility Criteria for designation as an FLAs; determine through analysis what defines "threatened" and "environmentally important forests"; and outline the State's project evaluation and prioritization procedures. The AON must be developed in consultation with SFSCC and approved by the State lead agency.

State lead agencies may utilize the services of land trusts or other entities in preparing the assessment. Information from existing sources may be used to prepare the AON, instead of initiating new studies that would duplicate existing data. Examples of appropriate sources include State Forest Resources Plans, State Comprehensive Outdoor Recreation Plans, growth management studies, State cultural site inventories, inventories of threatened and endangered species, and other State, regional or local plans, studies or reports.

The AON shall include relevant information about both public and private lands, address the issue of how best to maintain the integrity of forestlands for future generations, and address pertinent issues as identified by the State.

At a minimum, the AON must address the following as they relate to the purpose of the FLP:

1. Forest resources including:
   - o Aesthetic and scenic values;
   - o Fish and wildlife habitat;
   - o Minerals resource potential;
   - o Public recreation opportunities;
   - o Soil productivity;
   - o Forest products and timber management opportunities;
   - o Watershed values including water quality protection;
2. The present and future threat of conversion of forest areas to nonforest uses. States are responsible for defining the conversion threat(s);
3. Historic uses of forest areas, and trends and projected future uses of forest resources;
4. Current ownership patterns and size of tracts, and

trends and projected future ownership patterns;

5. Cultural resources that can be effectively protected;
6. Outstanding geological features;
7. Threatened and endangered species;
8. Other ecological values;
9. Public recreational opportunities;
10. Protected land in the State, to the extent practical, including Federal, State, and municipal lands and land trust organizations lands;
11. Issues identified by the SFSCC and in the public involvement process.

Using the above information the AON shall include the following:

1. Identification of applicable Eligibility Criteria;
2. Identification of specific FLA(s) for designation;
3. Specific goals and objectives to be accomplished by the FLP;
4. Process to be used by the State lead agency to evaluate and prioritize projects to be considered for inclusion in the FLP.

The project evaluation and prioritization process outlined in the AON should reflect the direction set forth in the CFAA to give priority to lands which can be effectively protected and managed, and which have important scenic or recreational values, riparian areas, fish and wildlife values including threatened and endangered species, or other ecological values. Traditional forest uses such as forest management activities, including timber management, and outdoor recreation opportunities are considered consistent with purposes of the FLP and are encouraged on FLP tracts when consistent with the State's AON and the conservation purposes for FLP tract acquisition. The prioritization process should implement a strategy that enhances existing protected forestlands or local and State conservation strategies as outlined in the AON.

The composition of the SFSCC is defined in Section 19(b) of the CFAA (16 U.S.C. 2113). States are encouraged to broaden this composition to include interests appropriate to benefit the FLP. This committee cooperates with the State lead agency in the preparation of the AON, identification of FLA Eligibility Criteria, the identification of proposed FLAs from which lands may be entered into the FLP, and recommendation of priority lands to be considered for enrollment in the Program.

Public participation and involvement in the AON preparation is a State responsibility. In the absence of established State procedures, NEPA may serve as an appropriate model for public involvement. The State lead agency will solicit involvement and comments on the AON from the public including State and local governments. The goals of public involvement include hearing concerns and views from interested and affected individuals and organizations, receiving new information, identifying and clarifying issues.

Based on the State-wide AON, the State lead agency, in consultation with the SFSCC, identifies specific geographic FLAs that meet the Eligibility Criteria, and recommends them to the FS for designation as of a FLA.

States are encouraged to cooperate in the identification of FLAs that cross State boundaries and to work together to coordinate acquisitions of lands or interests in lands that have complementary purposes. However, program implementation is undertaken by the individual States (State Grant Option) or by the FS (Federal Option).

The identification of proposed FLAs must include:

1. Location of each geographic area on a map and a

written description of the proposed FLA boundary;

2. Summary of the analysis used to identify the FLA and its consistency with the Eligibility Criteria;
3. Identification of important environmental values, and how they will be protected and conserved;
4. The conservation goals or objectives in each FLA
5. List of public benefits that will be derived from establishing each FLA;
6. Identification of the governmental entity or entities that may hold lands or interests in lands (State grant cption) or may be assigned management responsibilities for the lands and interests in lands enrolled in the program (Federal option); and
7. Documentation of the public involvement process and analysis of the issues raised.

Eligibility Criteria for Establishing Forest Legacy Areas (FLAs)

The CFAA directs the Secretary to establish Eligibility Criteria for the designation of FLAs, in consultation with the SFSCC. These criteria should be based upon the FLP purpose to protect environmentally important forest areas that are threatened by conversion to nonforest and be further developed through the AON.

FLA boundaries must encompass forestlands with significant environmental and other resource-based values. Areas may also include nonforested areas such as farms and villages if they are an integral part of the landscape and are within logical boundaries. Since FLA boundaries may not correspond to property boundaries, tracts located partially within the geographically defined FLA are eligible for the FLP, upon approval of a boundary adjustment by the FS Region/Area/IITF. Indian reservations and tribal lands may have important features on the forested landscape. Indian tribes and States are encouraged to collaborate and to consider only nontrust allotment lands for designation as, or inclusion within, a FLA. Other tribal lands are already protected through the trust relationship between the U.S. Department of the Interior and the tribe and are ineligible for the FLP.

States are responsible for determining what defines "threatened" and "environmentally important forest areas" in the State. However, environmentally important forest areas shall contain one or more of the following important public values, as defined by the States:

1. Timber and other forest commodities
2. Scenic resources;
3. Public recreation opportunities;
4. Riparian areas;
5. Fish and wildlife habitat;
6. Known threatened and endangered species;
7. Known cultural resources;
8. Other ecological values.

The FS, State or unit of State or local government may only acquire lands and interests in lands identified within a FLA under FLP authority on a willing seller/willing buyer basis.

AON and Amendment Approval

The State lead agency must submit the AON, including proposed FLAs and Eligibility Criteria, to the FS Region/Area/IITF. The FS Region/Area/IITF with input from the FS Washington Office reviews the AON and works with the State lead agency to complete the AON. Once finalized, the FS Washington Office forwards the AON to the Secretary for final approval. Final approval establishes the FLP for the State. AONs shall be periodically reviewed (at least at 5-year intervals) by the State lead agency, the FS Region/Area/

IITF, and the SFSCC to assess whether AON amendments or updates are necessary. The results of reviews will be documented by the State lead agency. AONs should be amended as needed.

The State lead agency may amend the AON to make significant changes or minor adjustments. Significant changes include modifications to their FLP, changes to the FLA Eligibility Criteria, or to add or delete a FLA. These changes need to be made in consultation with the SFSCC and with public involvement. FLAs and project evaluation criteria shall be of a scale and detail to effectively focus delivery of the FLP.

Significant Amendments to an AON may address the following:

1. Issues associated with maintaining the integrity of forestland and the proposed FLA specifically.
2. Revision, if any, of the FLA Eligibility Criteria.
3. Changes in policies or conditions that have occurred since the previous AON;
4. The identification of proposed FLA(s) and conservation goals or objectives associated with that FLA.

The Chief of the Forest Service, or designee, provides final approval of the Amendment of an AON authored by a State lead agency after consultation with SFSCC. In addition, the State lead agency may complete minor AON amendments, such as FLA boundary adjustment or project prioritization process. These minor changes need to be coordinated with the SFSCC, and need review and approval by the appropriate FS Region/Area/IITF.

Multi-State Identification of Forest Legacy Areas

States are encouraged to cooperate in the identification of FLA that cross State or Tribal boundaries and to work together to coordinate acquisitions of lands or interests in lands that have complementary purposes. States may elect to jointly use an existing or new multi-State or regional entity to identify FLAs or develop FLP projects that cross State boundaries. The entity must be a government-established organization, whose jurisdiction encompasses all or portions of the land area of the FLA States involved. However, program implementation is undertaken by the individual States (State Grant Option) or by the FS (Federal Option). The entity conducting a multi-State identification of FLAs is responsible for:

- o Obtaining approval from the appropriate States or Indian tribes for FLAs within their boundaries,
- o Cooperating with appropriate SFSCCs,
- o Obtaining public comments on the identification of FLAs, and
- o Complying with all other requirements of these guidelines.

Project Selection Process

The FS will conduct a project selection process to arrive at a prioritized national project list for consideration in the President's budget for the upcoming fiscal year. The project selection process and calendar of due date milestones are developed in consultation with the State lead agencies and FS Region/Area/IITF and communicated by the FS Washington Office. The FS will ensure that national evaluation and prioritization criteria are communicated to the States and in a timely manner so that submitted projects adhere to strategic goals and objectives of FLP. Project selection steps are:

Step 1: Release Project Selection Calendar with Due Dates

The project selection process and calendar of due date milestones are developed in consultation with the States and

FS Regions/Area/IITF and communicated by the FS Washington Office.

Step 2: State Project Prioritization and Submission

FLP project applications are accepted by the State lead agency as outlined in the State's AON. The SFSCC reviews and evaluates applications according to the criteria identified in the State's AON, authorizing statute, and other relevant direction and policy, and provides recommendations to the State lead agency. Projects approved and prioritized by the State lead agency are forwarded to the FS Region/Area/IITF for funding consideration. Only projects submitted through this process will be deemed eligible. Each State will prepare a list of projects and enter or update its list for submission to the FS via the Forest Legacy Information System or other means as requested.

Step 3: Forest Service Regional Review

FS Regions/Area/IITF will review submitted projects considering State priorities and national criteria. The purpose of this review is to improve project viability, facilitate the national project selection process and advance the strategic outcomes of the FLP. FS Regions/Area/IITF will submit projects to the FS Washington Office for funding consideration.

Step 4: National Review; Develop National Project List

The FS Washington Office will develop a prioritized national project list by convening a panel. There are 3 purposes of the panel;

(1) assure that all projects meet Congressional and Administration direction;

(2) assure that projects meet national program goals; and

(3) develop a National List of ranked projects.

The composition of the panel shall be developed annually in consultation between the State lead agencies and the FS, and will be representative of geographic regions. Project evaluation and ranking is based on the following national core criteria; project readiness will be considered as well as other evaluation considerations developed in consultation with State lead agencies and FS Regions/Area/IITF. The national core criteria are:

- o *Important:* The public benefits gained from the protection and management of the property including environmental values, and the economic and social aspects;
- o *Threatened:* Conversion to nonforest uses is likely or imminent and will result in a loss of forest values and public benefits; and
- o *Strategic:* Fits with a larger conservation plan, strategy, and initiative and enhances previous conservation investments.

States newly entering FLP will be given a "New-State start-up" preference for an initial FLP project. This is a placeholder for planning purposes and does not guarantee project funding. In order to receive the New State start-up project funds the State must have an approved AON and the project must meet national core criteria and the State's evaluation criteria and be submitted within the fiscal year that the placeholder is approved by Congress.

Step 5: Submit National FLP Project List to the Office of Management and Budget and to Congress

Each fiscal year, the FS Washington Office will submit a project list to the Office of Management and Budget for funding consideration in the President's budget. Once the President's budget has been completed, the FS Washington

Office will notify the appropriate House and Senate Committees and Subcommittees of the recommended projects for the upcoming Fiscal Year.

PROGRAM FUND CATEGORIES

Forest Legacy funds are allocated to one of three categories: Project Funds, Administration Funds, and AON Preparation Funds. FLP funds may not be used for monitoring and enforcement.

Project Funds

Project funds are those used to directly purchase lands or interests in land joining the FLP. Project funds may be expended by the State lead agency or the FS, as applicable, to cover transaction costs, including but not limited to: appraisals and appraisal review, land surveys, closing costs, establishing baseline information, title work, purchase of title insurance, conservation easement drafting, and other real estate transaction expenses for those tracts. Project funds may also be expended to facilitate donations of land or interests in lands to a qualified and willing donee for FLP purposes, by paying for expenses directly related to the donation, including but not limited to, land surveys, conservation easement drafting, title work, and establishing baseline information.

For an outright donation of a conservation easement or land, FLP program funds may not be used to pay for an appraisal. In the case of a partial donation of a conservation easement or land, an appraisal meeting Federal standards is required to determine the value of property. FLP funds may be used for appraisals on these partial donations. When Federal funds are used to purchase real property, including conservation easements, appraisal and acquisition work procedures must meet Federal standards.

Administration Funds

Administrative funds are the portion of funds used for day-to-day program management at all levels. Administration funds may be used for a variety of activities, including FLP program administration, personnel and overhead, and all activities identified as eligible uses of project funds to prepare projects and potential projects. Forest Legacy funds for administration shall be kept to a minimum. As a goal, all attempts should be made to keep administration funds under 15 percent of the total funds appropriated.

AON Preparation Funds

AON preparation funds may be made available to States to help defray the cost of preparing, or amending an AON.

Process for Allocating Funds to Forest Service Regions/ Area/IITF

Following passage of the annual appropriations bill, the FS Washington Office develops the Forest Legacy Program Direction and allocates funds to the FS Regions/Area/IITF for distribution. The allocation process differs for each fund category described below.

Allocation of Project Funds

Allocations to FS Regions/Area/IITF are based on the results of the national project selection process and the appropriations bill. Under the State grant option, FS Regions/ Area/IITF will award grants to States for specific, identified projects.

Allocation of Administration Funds

The FS Washington Office distributes administration funds to FS Regions/Area/IITF. Each FS Region/Area/IITF in consultation with the States requests these funds to meet their needs and the needs of the participating states in their

Region/Area/IITF. Administration funds are also used by the FS Washington Office to fund program management functions. Administration funds will be granted to States under the State grant option separately from project funds.

Allocation of AON Preparation Funds

The FS Washington Office distributes AON preparation funds to the States by way of FS Regions/Area/IITF. These funds are requested by FS Regions/Area/IITF to meet the needs of their States to develop new AONs or amendments.

Redirection and Reprogramming of Funds

Due to the nature of real estate transactions, FLP projects may change in scope, cost or fail completely. These changes can result in unspent or excess funds for some projects while others may need additional funding to bring them to completion.

In order to maximize the efficient and effective use of FLP project funding, the FS will either redirect or request reprogramming of funds. Redirection is a shift of funds from one congressionally approved project to one or more other congressionally approved project(s). Reprogramming is a shift of funds that exceeds an increase or decrease of 10% per project not to exceed $500,000 to an existing project, or shifting of any amount of funds to a project not previously approved by Congress.

Regional Redirection Process

FS Regions/Area/IITF may redirect up to an increase or decrease of 10% per project not to exceed $200,000 of project funds that are excess or unspent from one project to one or more other Congressionally approved project(s) within the FS Region/Area/IITF which is underfunded and where there is a substantiated need (e.g. loss of other funding sources, appraisal documenting increased cost, etc.) to bring the project to completion.

Project funds over $200,000, or those that cannot be redirected by the FS Region/Area/IITF, will be released for the national process. FS Regions/Area/IITF will notify the FS Washington Office before a redirection takes place and report these actions periodically. All funds from failed projects will be released for the national process.

National Redirection Process

The FS Washington Office, through consultation with FS Regions/Area/IITF, may redirect up to an increase or decrease of 10% per project not to exceed $500,000 of project funds that are excess or unspent from one project to one or more congressionally approved project(s) which is underfunded and where there is a substantiated need (e.g. loss of other funding sources, appraisal documenting increased cost, etc.) to bring the project to completion. In addition, when funds have not been spent or contractually obligated within two years of receipt of funds, they revert to the FS Washington Office via the appropriate FS Region/Area/IITF. The FS Washington Office will:

1. Assess the extent of unspent or returned funds on a periodic basis;
2. Facilitate selection and funding of underfunded projects not addressed by the regional process or between FS Regions/Area/IITF; and
3. Notify Appropriations Subcommittees of any redirection action taken by the FS FS Regions/Area/IITF or FS Washington Officc.

National Reprogramming Process

The FS Washington Office, through consultation with FS Regions/Area/IITF, may request reprogramming by the Appropriations Subcommittees of unspent or returned funds to a project that requires more than an increase or decrease of 10% per project not to exceed $500,000 to complete. In addition, the FS Washington Office may request reprogramming by the Appropriations Subcommittees of unspent or returned funds to a project not previously approved by Congress. The FS Washington Office will:

1. Determine the funds available for reprogramming on a periodic basis.
2. Identify underfunded projects that cannot be addressed through the Regional Redirection Process and determine the priority for reprogramming.
3. Recommend reprogramming to fund projects from the National Project List next in sequence in priority ranking to the extent practicable.
4. Submit reprogramming requests to the Appropriations Subcommittees for approval.
5. Allocate funds to projects approved for reprogramming.

FLP Cost Share Requirements

The CFAA directs that, to the extent practicable, the maximum Federal contribution for total program costs may not exceed 75 percent. To assure program-wide cost share goals are met, each project budget must include a minimum nonfederal contribution of 25 percent. This nonfederal cost-share must meet Forest Legacy purposes. It may consist of:

(1) the value of land, or interest in land, dedicated to the FLP that is not paid for by the Federal government;

(2) nonfederal costs associated with program implementation; and

(3) other nonfederal costs associated with a grant or other agreement that meets FLP purpose.

The nonfederal cost-share must be documented, and in

the case of a grant, must meet the timing, terms, and conditions of the grant. The cost-share can occur at any phase of the FLP including planning, developing future projects, acquisition, capital improvement, management, or administrative activities. When a grant is involved, the cost-share must occur within the life of a grant and meet all grant requirements. Federal requirements identify the grant period as beginning when the grant is formally awarded and ends after two years to ensure that the federal funds are spent promptly. However, a grant may receive a maximum extension to five years. Allowable costs shall be determined in accordance with the 7 CFR 3016, "Uniform Administrative Requirements for Grants and Cooperative Agreements to State and Local Governments," and any amendments to this regulation.

Donations of land or interests in land must be documented to count as part of the nonfederal cost-share. The title does not need to be transferred to the State or federal government in order for the donation to qualify as cost share. However, if in the future, the donated lands are conveyed or the rights or title are modified in a way that is inconsistent with the purposes of the FLP then the State must restore the cost share value dedicated in the grant agreement. The value of donations may be included as part of the nonfederal cost-share if all of the following are met:

1. The donation contributes to the objectives and priorities of the State FLP as set forth in the AON;
2. All or part of the tract being donated must be within the boundaries of an FLA, and may include National Park, National Forest, National Wildlife Refuge, or other Federal land boundary, or within the boundaries of an area designated through an analogous State program with goals compatible with the FLP and be within an FLA;

3. The donor documents their desire that value of the interests may be used as cost share for the FLP project;
4. The donation of land or an interest in land must contain perpetual covenants to assure that the tract will be managed in a manner compatible with the goals for which the FLA was established;
5. The donee (holder of donated rights) is a unit of government or a non-profit conservation organization (land trust) that meets the eligibility requirements for holding a conservation easement established by the Internal Revenue Service and has as its purpose the management of lands or interests in land consistent with FLP purposes;
6. If the donation is in the form of a conservation easement then the deed needs to contain a provision that directs all of the easement holder's proceeds from a subsequent sale or exchange of interests in land be used in a manner consistent with the conservation purposes identified for the subject interests in lands;
7. The respective portion of the donation must not have been previously credited towards any Federal program's nonfederal cost share; and
8. The State lead agency approves the donation as contributing to the cost-share.

GENETIC DIVERSITY, BREEDING AND UTILIZATION OF CITRUS FRUITS

Citrus fruits are the third largest of the fruits grown in India with an estimated production of 2979.11 thousand m.t. from an area of 369.56 thousand ha.

The country is the home of many citrus species, namely, Citrus inchangensis, C. latipes, C. macroptera, C. assamensis,

C. aurantium, C. jambhiri, Citrus limonia, C. karna, C. pennivesiculata and C. maderaspatana. The diverse ecogeographical distribution and the occurrence of spontaneous mutations and natural hybridization have given rise to a wide range of variability in citrus and related genera. Genetic resources of citrus in India have been well reviewed (Singh and Chadha 1993; Chadha and Singh 1996).

In India, more than 500 citrus varieties are available of which, mandarin, sweet orange and acid lime are commercially grown with regional preference considering the agroecological conditions, resulting in their increased area and production. Increase in production over the base year of 1961 is over 242.59% with an annual growth rate of 7.82%. Citrus fruits thus occupy 11.53% of the total area under fruits in India. However, genetic base of cultivars is very narrow and enrichment of germplasm needs greater attention.

DISTRIBUTION OF DIVERSITY

There are three major centres of diversity. The types found include the papeda, pummelo and their hybrids, Citrus indica and many types of citron, lemons and mandarins. These include interesting types like Nemerutenga, Soh-Systeng, a sour fruit similar to the sweet lime and Soh-Siem, a sour mandarin. Diversity of this region is well studied and described by Bhattacharya and Dutta (1956). In South India, the indigenous types include Gajanimma or Baduvapuli, Kichli and some wild mandarin types, viz., Kodakithuli, Billikichili, Nakoor lemon, Mole puli (sour orange type), etc. In western India, at the foot of the Himalaya, the hill lemon (galgal) and Attani are common.

Citron, Citron lemon, Karna Khatta, Rough lemon, Rangpur lime, acid lime, hybrid pummelo and various types of mandarins are found all over the country. Wild types or plants growing without any attention are largely found in the foot hill regions of northwestern India, in the north-east, in south in Malabar hills and in the Western Ghats under varying ecological conditions. Distribution of indigenous cultivars is presented in Table below.

CHARACTERIZATION

Characterization of citrus genetic resources was first attempted by Bonavia (1890) who described the citrus germplasm available in India and Lushington (1910) described the Citrus species of north-eastern region. Other workers have also described the Citrus germplasm. These studies were based on taxonomic observation of herbarium specimen. Systematic characterization and documentation of citrus in north-eastern India was done by Bhattacharya and Dutta (1956) following the descriptors of Swingle and studying minutest details of the samples. Hodgson et al. (1963a,b,c), Singh (1967, 1977), Singh and Singh (1967, 1968), and Singh and Nath (1969) also described Citrus species. However, these morphological descriptions could not remove the confusion in classification but on the contrary, the divergent views complicated the Citrus taxonomy.

Further, the chemotaxonomic approaches have been used to confirm the characterization. With use of pictorial scatter diagram for characterization, Singh and Singh (1983) used numerical classification using 145 characters. For assessing diversity, electrophoresis of proteins and enzymes has proved to be an authentic tool in characterization of citrus germplasm at gene level. Biochemical markers may help in eliminating duplicates in genepool resulting from variations in phenotypic expression due to environmental conditions.

EVALUATION

Citrus genepool has been evaluated for agri-horticultural characteristics (Chadha and Singh 1990; Singh 1990). The evaluation at different locations has helped to identify cultivars/rootstocks for different agroclimatic situations as well as identification of suitable rootstocks. Variability in mandarin and acid lime was studied to understand the genotype × environment interaction (Singh et al. 1980; Chakrawar and Jature 1980; Prasad and Rao 1989).

Several citrus species types have been evaluated for their suitabilty as rootstocks, cultivars resistant to insect pest and disease and for their processing qualities. So, by effective evaluation, it has been possible to identify and domesticate the exotic and indigenous collection for commercial adoption. Kinnow mandarin growing commercially in Punjab is one of the best example of successful introduction.

DOCUMENTATION

Although documentation of genetic resources of Citrus has been done (Bhattacharya and Dutta 1956; Singh 1967; Singh and Singh 1977; Srivastava and Singh 1977; Iyer and Dinesh 1987; Singh and Uma 1995) for better accessibility of information, database with central monitoring system is still lacking.

UTILIZATION

There are ample evidences to show that genetic resources of citrus have been well utilized (Singh and Chadha 1993). In fact, the germplasm collected from exotic or indigenous sources have been the basis of commercial cultivation in different citrus growing regions in the country. By and large, the commercial cultivars are the result of selection from indigenous diversity.

Owing to inherent problems of breeding, only limited attempts have been made for the transfer of desired genes through hybridization. However, natural variability existing in commercial cultivars or related species have been utilized. Extensive survey of acid lime orchards in Maharashtra state resulted in selection of two commercial cultivars viz. Pramalini and Vikram which are superior to existing acid lime cultivars. Ten clones of acid lime were assembled and evaluated at Periyakulam for 15 years, which resulted in selection of superior types and release for commercial cultivation. At Rahuri, 150 accessions collected from different lime growing regions were evaluated which has enabled to identify 5 superior types with respect to yield, fruit quality, bearing period and disease resistance. One of the accessions, Selection-49 has been released for growing in Maharashtra as Sai Sarbati.

Exotic germplasm viz. Kinnow, Seedless lime, Jaffa, Valencia Late, Marsh and Grape fruit have become important commercial cultivars in different citrus growing regions (Singh and Chadha 1993). Unshiu introduced from Japan has also promise with its short duration and dwarf canopy. Indigenous collection, Nepali Oblong lemon has proved to be a best donor for imparting resistance to canker in hybridization programme. The efforts have resulted in development of acid lime having resistance to citrus canker, a serious problem of acid lime (Prasad M.B.N.V. 1997, personal communication).

The reviews of rootstock situation in India suggest that Rangpur lime is the most promising for mandarin and sweet orange in central and South India. Performance of Cleopatra mandarin has also been highly satisfactory. However in Punjab, Jatti Khatti (C. jambhiri) and Rangpur lime for Kinnow; Rangpur lime and Cleopatra mandarin for Blood Red, Jatti Khatti and Cleopatra mandarin for Jaffa have shown promise (Dhatt and Singh 1993). A 25-year rootstock trial suggested Belladakithuli (C. maderaspatna) as the best for Coorg mandarin with respect to fruit yield (Srivastava et al. 1983) but was not recommended due to bottleneck growth at graft union. Thus, this rootstock needs further testing. Feronia limonia proved to be highly dwarfing and precocious at Shrirampur and Tirupati, and is thus of interest for high density planting. Since, rootstocks currently employed have one or other defect, work to develop ideal rootstock having resistance to root rot, nematode, salinity with good agronomic characters, by citrus breeding was started at Chethalli wherein trifoliate orange was used for imparting resistance (Singh and Chadha 1993).

Trifoliate orange has been found to be a donor for the resistance to root rot (Iyer et al. 1983; Prasad and Rao 1983) and nematode (Reddy et al. 1987). Several hybrids having tolerance to diseases and nematode also have promising growth characters. These hybrids are being tested for their suitability as rootstock for mandarin and sweet orange.

TAXONOMIC STATUS

The commonly grown citrus species belong to three genera, Citrus, Fortunella and Poncirus. All these genera are closely related and have inter-generic fertility. These genera are grouped under the subtribe Citrinae, tribe Citreae, sub-family Aurantioideae and family Rutaceae.

Divergent views on classification of Citrus have been expressed. Swingle and Reece (1967) recognised 16 species, while Tanaka (1954) proposed as many as 145 and later 159 (Tanaka 1961). Tanaka (1961) classified the forms of Citrus into two sub-genera, eight sections, fifteen sub-sections, nine groups, two sub-groups, two micro-groups and 159 species. Swingle's classification had recognized the sub-genera Papeda (containing six species) and Citrus (formerly Eucitrus) with ten species (Swingle and Reece 1967). Bhattacharya and Dutta (1956) published a monograph on classification based on the study of a large number of specimens of citrus fruits collected during 1938-48 from wide range of soil and climatic conditions up to 2000 m altitude in Assam. They followed the classification of Swingle in a general way, but accepted C. nobilis Lour., C. karna Raf., C. limetta (Risso) Lush., C. jambhiri Lush., C. megaloxycarpa Lush, and species of Tanaka that were not accepted by Swingle (1948). Their classification found C. pennivesiculata Tanaka, and C. limonia Osbeck to be invalid. They named a new species, C. assamensis Dutta and Bhattacharya, which is known as Adajamir in Assam. Validity of this species was, however, questioned by Singh and Nath (1969) as it differs from the C. pennivesiculata Tanaka essentially in the nature of the essential oil of peel. This was identified by Tanaka as C. medica var. alata (Singh and Nath 1960).

Subsequently, large exotic introductions were studied by Hodgson et al. (1963a,b,c) who recognized twenty three species cf Tanaka. Singh (1967) compared characters used by Swingle and Tanaka, and proposed a key to the identification of Citrus compromising the two systems and recognized C. maderaspatana Tan. (kichli), C. unshiu Tan., C. deliciosa, C. pennivesiculata Hort. ex Tan., C. semiperflorens and C. limonia Osbeck as valid species besides a new species, believed to be a complex natural hybrid of acid lime (C. aurantifolia Swingle) and papeda group parentage. Being distinct with least resemblance to any other form of citrus(it was designated as Citrus nakoor D. Singh sp. novo).

Considering the complexity of classification based on morphological data which is difficult to comprehend in terms of pattern of variation arising due to natural hybridization and mutation, numerical taxonomy has been attempted (Singh and Singh 1968; Prasad 1995). It has been found eminently suitable in assessing the similarities between different species for their classification into groups and their phylogenetic relationship. Use of pictorial scatter diagram in clarifying the position of hybrids, probably of bispecific origin, is also advocated. Polygonal method of presentation (after Hutchinson) of morphological data was found useful in characterising different citrus types.

To confirm this, chemical constituents have been proposed as additional tools. By combination of modified almen test, ammonium molybdate test and ferric chloride test, it was possible to distinguish between Citrus limon and C. jambhiri, C. limettioides and C. aurantifolia; C. grandis and C. paradisi. Usefulness of flavonoid composition of leaves, bark and ripe fruit peel in identifying the taxa of doubtful origin was also emphasised. Chemotaxonomic studies were employed to study the variability in lemon, mandarin and related citrus types. Study of 145 morphological characters, affinity relationships, chemical data and breeding behaviour favoured the contention that only three species, C. medica, C. grandis and C. reticulata with a sexual reproduction system are in Eucitrus group.

Prasad (1995) studied 71 cultivars belonging to 31 species for genetic diversity and genotype affinity using D2 analysis and leaf peroxidase isozyme. The genotypes exhibited wide variability for morphological characters. Numerical analysis and polygraph suggested some affinity at intra-specific level. Very narrow relationship between Rangpur lime, to very wide relationship between pummelo and sour orange revealed the broad nature of genetic divergence in citrus. Based on minimum generalized distances, the 71 genotypes fell into four distinct clusters. Leaf peroxidase isozyme banding pattern proved the genetical differences and distinct genotypes. The isozymes resolved the differences between seemingly similar types of citrus, particularly at intra-specific level. He also noted the close affinity among mandarin orange, sweet orange, acid lime and Rangpur lime inspite of enormous diversity.

CONSERVATION

Germplasm is maintained under different agroclimatic conditions at Central Horticultural Experiment Station,

Chethalli, Karnataka; Indian Institute of Horticultural Research, Bangalore, Karnataka; Regional Fruit Research Station, Abohar, Punjab; Horticultural Experiment Station, Bhatinda, Punjab; Division of Fruits and Horticultural Technology, IARI, New Delhi; Department of Horticulture, Rahuri, Maharashtra; Citrus Improvement Project, Tirupati, Andhra Pradesh; Citrus Experiment Station, Katol, Maharashtra; Horticultural Experiment Station, Periyakulam, Tamil Nadu; Citrus Experiment Station, Tinsukia, Assam and others. At these centres, germplasm is maintained as working collections.

Citrus germplasm is conserved in field genebanks at different locations (Srivastava and Singh 1977; Iyer and Dinesh 1987; Singh and Uma 1995). Though substantial diversity is conserved in field genebanks, it does not contain total representation of the variability available in citrus.

Many of the germplasm collected by Bhattacharya and Dutta (1956) from north-eastern India are currently not available in any of the collections. At the same time, many of the germplasm are safely duplicated at few centres owing to the fact that germplasm of one region behaves differently at other region.

IN VITRO CONSERVATION

Maintenance of citrus germplasm in field genebank requires large area and funds, and plants are also affected by insect pests and diseases. Therefore, in vitro conservation of the citrus germplasm may require attention.

In vitro conservation technique is also used for revitalization of citrus germplasm affected by virus and virus-like diseases as meristem culture technique is found to eliminate many of the viruses. However, to conserve the germplasm in vitro, there is need for the research with respect to genetic stability and duration of storage, etc.

Proliferation in callus was observed even after several

years by sub-culturing in medium supplemented with 5 per cent sucrose alongwith 2,4- D, NAA and kinetin each of 0.25 mg l-1.

In vitro root tip cultures continued to retain the ability to form shoot buds which were maintained at 25°C for 8-10 months and regenerated to plants.

Table : Citrus germplasm maintained at different centres in India

Germplasm	Cultivars/Accessions		
	1977	1987	1995
Sweet orange (C. sinensis Osbeck)	132	110	107
Mandarins (C. reticulata Blanco)	58	58	68
Mandarin hybrids	22	11	26
Tangeloes	24	17	14
Satsuma mandarin (C. unshiu Marc.)	20	10	15
Small fruited mandarin	12	14	14
Acid lime (C. aurantifolia Swingle)	30	40	46
Seedless lime (C. latifolia Tan.)	4	2	2
Rangpur lime (C. limonia Osbeck)	27	30	28
Lemon (C. limon Burm:)	63	57	42
Grapefruit (C. paradisi Macf.)	45	27	33
Pummelo (C. grandis L. Osbeck)	36	21	19
Rough lemon (C. jambhiri Lush.)	55	55	60
Sour orange (C. aurantium L.)	23	17	18
Sweet lime (C. limettioides Tan.)	11	6	2
Citron (C. medica L.)	9	8	15
Other citrus species and their cultivars	59	43	83
Natural hybrids	13	22	17
Poncirus trifoliata (trifoliate orange)	21	22	18
Trifoliate hybrids	36	41	30
Related genera	15	15	10

Citrus Breeding/Improvement

Despite large variability that exists in citrus fruits, a cultivar having the ideal plant type, precocity, regular bearing, attractive and good quality fruits with resistance to pests and diseases is still lacking and the varietal situation remains more or less static.

Since the production of citrus fruits are governed by suitable stonic combinations, root-stock trials have attracted major attention. Till todate, more than 159 varieties in different species of citrus have been screened for their suitability as rootstock to different citrus fruits at various locations which has enabled to identify more than 10 cultivars suitable as rootstocks for different scions under varying agro-ecological conditions of which Rangpur lime and Cleopatra mandarin are being commercially adopted (Chadha and Singh 1990; Singh 1993, 1994), but none of the cultivars have all the desirable characters.

Table : Characteristics of few selected rootstocks

Rootstock	Horticultural yield	Reaction to diseases								
		Quality	Plant vigour	Root rot	Citrus nematode	Tristeza	Exocortis	Salt	Drought	Root system
Rangpur lime	G	M	G	MT	S	R	S	R	R	D
Marmalade orange	G	M	G	MT	MT	R	S	R	R	D
Rough lemon	G	L	G	S	S	R	R	T	T	D
Cleopatra mandarin	M	M	M	T	S	R	R	MT	D	M
Dout orange	G	G	M	R	T	HS	T	T	MT	D
Sweet lime	M	M	G	S	MT	S	S	S	S	M
Trifoliate orange	L	M	L	R	R	R	H	HS	HS	Sh
Troyer Citrange	M	G	M	MT	T	MT	S	HS	HS	Sh
Carrizo Citrange	M	G	M	M MT	T	MT	S	HS	HS	Sh
Sweet orange	G	G	M	HS	HS	MT	R	S	S	M
Kama Khatta	G	M	M	S	MT	-	T	T	S	D
Nasnaran	M	G	M	S	MT	T	R	HT	S	M

Source: Chadha and Singh 1993.

G: Good, M: Moderate, L: Poor or low, R: Resistant, T: Tolerant, MT: Moderately tolerant,

S: Susceptible, HS: Highly susceptible, D: Deep, M: Medium, Sh: **Shallow**.

For instance, the variety which has excellent quality of fruits is highly susceptible to diseases. Dwarfness, resistance to insect pests and diseases are lacking.

The root-stock which is resistant to root rot and nematode has high susceptibility to drought and salinity. Thus, need for combining all these characters through interspecific crosses was started at Central Horticultural Experiment Station, Chethalli, Coorg (Iyer et al. 1983) in 1972 as well as at other locations. At other locations, breeding for rootstock was phased out and the programme started at Chethalli was shifted to Bangalore in 1978.

This programme has resulted in identification of some promising hybrids. Breeding work initiated at Chethalli for scion improvement has not yet yielded any hybrid. By and large, improvement work is confined to introduction of improved material from indigenous and exotic sources and its evaluation under different climatic conditions. Recently, selection of superior clones in commercial cultivars has received attention.

An important goal in breeding scion and rootstock is recombination of desired characters into one. The selection of parents depends upon the objectives and characters to be combined. However, it is not certain that such characters will regularly appear in hybrids. At the same time, undesirable character, which are not phenotypically seen, may appear in F1 hybrids. The cultivars which have less polyembryony should be selected as parents. It is also desirable that pollen parent should have marker characters for identification of hybrids.

Floral biology of most of citrus types are well studied and reviewed for successful hybridization (Iyer et al. 1983). Till date, no success has been achieved in recombination of characters in scion cultivars. However, promising results have been obtained in recombination of resistant character in rootstock varieties (Iyer et al. 1983). None of the rootstocks presently available have all the desirable characters to make an ideal rootstock for citrus.

Thus, systematic breeding programme for developing an ideal rootstock variety was started first in 1972 at Central Horticultural Experimental Station (CHES), Chethalli, wherein large number of crosses were affected by synchronizing the flowering, storage of pollen and identification of hybrids using marker gene (Iyer et al. 1983).

Simultaneously, rootstock breeding was started at Rahuri and Tirupati but these programmes were phased out subsequently. The breeding programme started at Chethalli was continued at Bangalore and from the screening of hybrids for the growth and resistance to Phytophthora and nematodes, three promising hybrids have been selected. These hybrids are being used for mandarin and sweet oranges to evaluate their performance as rootstock at different locations. Rootstock breeding programme has also been started at NRC for citrus at Nagpur.

Breeding of acid lime for resistance to canker was started at Tirupati and Parbhani in 1976. Large number of crosses were made using different combinations but none of these hybrids were found promising, and the programme was phased out after about 10 years. Systematic approach in this direction was made at IIHR, Bangalore using Nepali Oblong as major source of resistance. Large number of hybrids are in advanced stage of evaluation. One of the hybrid from these crosses has comparable quality added with resistance to canker (Prasad M.B.N.V. 1997, personal communication).

BIOTECHNOLOGICAL EMPHASIS

Isozymes banding pattern found useful in characterization of genepool of many fruit crops was attempted by Prasad (1995). The leaf peroxidase isozyme banding pattern studied in 71 genotype exhibited highly polymorphic bands. Fast moving isozyme (Per.-1) had one band, followed by Per.-3 with two bands.

The next zone, Per.-3 with 11 bands, Per.-4 with 5 bands and slowest moving Per.-5 only one band. Per.-5 was present in all the 71 genotypes and did not contribute to any variability while Per.- 1 was specific to Nepali Oblong lemon. Per.-2 was found in Molepuli, Keer sour and C. amblycarpa.

Thus, most of variability in genotype was found for Per.-3 and Per.-4. Among the mandarin, Lahore local revealed its distinct identity with presence of additional band. Pummelo had five bands at three loci. Grapefruit, believed to be a natural hybrid between orange and pummelo, also exhibited affinity as observed through banding pattern. Rough lemon has proved to be a distinct genotype than other lemon. Distinct banding pattern was observed for different strains of Rangpur lime and total variability in their group was considered to belong to three genotypes. Sour orange exhibited variation in banding pattern. Nakoor lemon differed genetically by exhibiting their distinct peroxidase banding pattern.

The studies revealed that except Rangpur lime where some member were found to have phenotypic differences, all other genotype were genetically different.

DNA markers have been developed and applied for the identification and classification of citrus germplasm. These markers include restriction fragment length polymorphism (RFLP), random amplified polymorphic DNA (RAPD) and other finger printings. DNA markers as well as isozyme markers give valuable information on the phylogenetic relationship of Citrus species. DNA marker can also be used to identify minor variation generated by setting or hybridization to be identified.

5

UTILIZATION OF FOREST PRODUCTS

INTRODUCTION

The pattern of utilization of forest resources has been influenced by growing rural population, with limited income opportunities and the related widespread poverty and migration of landless people in the forest areas. This pattern has intensified agricultural cultivation and inappropriate exploitation of forest resources, resulting in depletion and degradation of the natural asset. However, local communities, living in or adjacent to protected areas, have continued to use the forest resources extensively in order to supplement their diets and household incomes, notably during certain periods in the year (Arnold & Perez, 2001). Forests are also the dominant source of household energy for cooking, construction materials, animal fodder and traditional

medicines. This paper examines the utilization of forest products in the buffer zone of Bach Ma National Park in Central Vietnam. In particular, it seeks to quantify the exploitation of forest products so that productive use values can be calculated. The establishment of the National Park in 1991 has undermined the legal exploitation of the natural resource base for the local users, but it has also enhanced and ensured environmental services as a common good for the buffer zone communities. Hence, current study also intends to assess local people's awareness of the non-marketable functions and services Bach Ma National Park provides.

STUDY SITE

Bach Ma National Park is situated in Central Vietnam, 40km south-east of the old imperial city of Hue. The area of the park is located in Thua Thien Hue Province, within the latitude of 16°05' - 16°15' Nand the longitude of 107°43' – 107°53' E. The park covers a total area of about 22,031ha which is managed with a minimum of human interference. Bach Ma National Park is identified in Vietnam's Biodiversity Action Plan as one of the last remaining Type A primary forests which should receive the highest priority for protection due to its biodiversity value. The National Park was created in 1991 with the aim to conserve the only green transect left in Vietnam, stretching from the South China Sea to the border with Laos.

The dominant habitats are tropical evergreen monsoon forest in the lower areas and subtropical evergreen monsoon forest at altitudes higher than 900m. The flora of Bach Ma includes at least 1,400 species. This represents around 19percent of the entire flora of Vietnam in only 0.07 percent of Vietnam's total land area. The park is located within the transition zone of northern (Sino-Himalayan, Indo-Burmese) and southern (Malesian) floras and is regarded to be an important 'Floristic Biodiversity Centre' for Indochina (Tran & Ziegler 2001). The protected area is surrounded by a buffer zone of approximately 21,300 hectares. Around 65,000 people in 12,285 households are estimated living there. Population density in the buffer zone is high (158 inhabitants per km and is likely to increase within the near future.

Local communities largely cultivate wet rice and the society is characterized by a high dependency upon agriculture. The socio-economic situation in the buffer zone of Bach Ma National Park is in many ways similar to other protected areas throughout Vietnam: the average income per head is low at around 250 USD per year, equivalent to 70 percent of the national income per head, and 40 percent of all households are classified as poor. Furthermore, the area provides difficult conditions to achieve high agricultural output and is blighted by natural disasters. The severe flood in 1999, for example, has reduced the growth rate in the agricultural sector considerably to -3.9 percent in that year (Phu Loc Statistical Office 2001). Consequently, without alternative economic incentives to agriculture, many households continue to use and commercialize illegal forest products, such as timber, firewood, and non-wood forest products.

METHODOLOGY

Field data were collected in the period from April to July 2001 in the buffer zone of Bach Ma National Park. Four villages belonging to two communes whose inhabitants are largely dominated by the Kinh ethnic group were chosen as a representative test sample. All villages are located in close proximity to the boundary of the protected area, and were founded after the liberation of South Vietnam in 1975 when the Vietnamese government launched a large resettlement programme in the aftermath of the American War. In total, 70 households were interviewed by using a semi-structured questionnaire.

Questions focused on the use of timber, non-timber and non-wood forest products but non- consumptive forest services were also investigated. Information was gathered on the utilization of forest products, seasons and frequencies of entries into the protected area for collection purposes, and a rough estimate was given on the quantity of forest products harvested and collected. The interviewees were also asked to what extent they consider the forest as important to support their livelihoods and whether they could imagine living without the protected forest.

The obtained list of products was categorized into timber, firewood, and non-wood forestproducts (NWFP). The definition of NWFP followed the classification given by FAO (1999) where NWTP are all biological materials other than wood which are extracted from natural forests for human use as well as services derived from forests and allied land uses. In practice, collecting sound data on forest use from interviews turned out to be difficult: interviewees were often reserved and hesitated to answer detailed questions, as they were generally aware of the regulations of the government of Vietnam on collection of biological materials in National Parks (Mai & Tran, 1996). Those who invade and clear the forest or collect forest products risk several punishments. These punishments depend on the kind of extraction and vary from confiscation of tools and material collected for smaller violations to heavy fines or imprisonment for extraction of timber, high value plants and animals susceptible to endangerment. Hence, additional information on the utilization of forest products was also obtained by direct observation in the villages and in a major market located in the buffer zone.

RESULTS

The surveyed households use a variety of forest products and the survey also indicates that the use of forest products is particularly crucial for marginal households as it provides a prompt income with no significant material investment. Furthermore, the dependency upon forest resources is accentuated in situations by lack of money or actual food shortage.

Forest management expenditures, which are made in the expectation of creating benefits at some time in the future, are best viewed as investments. Although some foresters are reluctant to take this perspective, there are compelling reasons for doing so.

This opens up opportunities for using standard investment analysis techniques to rank projects competing for scarce funding. The use of these techniques lets forest managers evaluate alternative strategies and select the one which will contribute the most to the achievement of management goals. This technical note is intended to provide forest resource managers with an introduction to benefit-cost analysis and its potential use in forest management. We present a basic introduction to some important concepts of benefit-cost analysis, then discusses the nature and valuation of benefits and costs commonly found in forestry, and finally concludes with an illustrative example.

NORTHEAST SCIENCE & TECHNOLOGY

The basic purpose of benefit-cost analysis (BCA) is to assess the economic attractiveness of an action or project. A single project may be assessed to see whether it is profitable or meets a minimum profitability standard. BCA is also commonly used to rank alternate projects and select the most attractive option. Project assessment is based on a comparison of the costs of the project against the expected benefits. There are several ways in which the costs and benefits can be compared, but all of these measures share certain key characteristics.

In forestry projects, the benefits and costs are often incurred over a long period of time and they have to be discounted to allow for a consistent comparison. Discounting converts the value of future benefits and costs to today's dollar values. For someone with a discount rate of r percent per annum, today's value of $100 payable to that person in one year can be calculated as:

Present Value = $100/(1+r/100)

When r is ten percent, the present value equals \$90.91. Looked at it in another way, \$90.91 today can be seen as the present value of \$100 payable one year in the future because \$90.91 invested at ten percent will yield \$100 in one year. Discounting is necessary because it is not accurate to treat equally dollars that are received or paid out at different times.

There are three measures commonly used to compare the costs and benefits. The approach that most economists favour is simply taking the difference between the total benefits and total costs. The result is known as the present net worth (PNW) or present value of a project or action. It is the net gain of the action. The formula may be written as follows:

(1) PNW = Benefits-Costs

When one is comparing alternate projects, the project with the highest PNW is the most desirable because it provides the maximum net gain. A project with a PNW just breaks even and a negative PNW implies a net loss from the project or action.

An alternative ranking approach is to calculate the value of the benefits derived per cost dollar. This is known as the benefit/cost ratio (BCR), and may be written as:

(2) BCR = Benefits/Costs

The project with the highest BCR yields "the biggest bang for the buck" and is the most desirable. A break-even project has a BCR of one and a money-losing project has a BCR of less than one.

The third common ranking criterion is the internal rate of return (IRR), which is defined as the rate of discount that produces a PNW of zero. Again the highest IRR is associated with the best project.

Of these three approaches, the IRR will not be explored further, since it can be difficult to calculate when the investment period is very long, as is often the case in forestry. The PNW is often the best criterion for ranking projects, but the BCR is a measure that can useful in identifying forest management alternatives that lead to higher PNWs.

Benefit-cost analysis can be used to evaluate forest management at various scales. At the smallest scale, BCA can be used to compare regeneration alternatives for a particular site. At a larger scale, alternative regeneration strategies for a management unit might be compared. For example, there is much debate over the degree to which intensive forest management should be practised and BCA can be used to compare the benefits of pursuing strategies based on different management intensities. BCA can also be used to allocate funding among management units, or competing projects.

Identifying and measuring the benefits and costs is generally easier on smaller projects. The most common benefit of forest management is an increased volume of timber, either per hectare or from a management unit. Usually, the application of management changes the species that are found, especially on reforested areas, and it may lead to the production of higher quality stems. Management may also lead to increased population levels of wildlife species, increased or improved recreational opportunities, better regulation of water flow, and improve the aesthetics of an area.

Typical costs of forest management are expenses for silviculture, pest and fire protection, access and harvesting roads, and overhead. If a project is difficult to finance or can be expected to lead to cash flow problems, the cost of financing should also be included. Other costs or disbenefits may be incurred if wildlife populations or recreational opportunities are reduced, if the aesthetic quality of an area is reduced, or if water regulation is disrupted. Measuring the size of these disbenefits is often difficult but is necessary if they are to be valued. Some estimation will usually be required. Clearly, poorly planned and executed forest management activities generate costs that can be avoided.

The analyst must also select a discount rate. There are two chief options: the "real" discount rate (which has inflation netted out) and the "nominal" discount rate, which includes inflation. The advantage of excluding inflation is that there is no need to estimate future inflation levels and adjust future costs and benefits accordingly.

That leaves the analyst with the task of selecting an appropriate real interest rate. If the prime interest rate is denoted as r, the inflation rate as p, and the real interest rate as i, the relationship between these variables is given by:

(3) $(1 + r) = (1 + i) * (1 + p)$

Setting $r = 0.07$ (i.e seven percent) and $p = 0.02$ (i.e. two percent) results in a real rate of interest equal to 0.049 (4.9 percent). This is in the middle of the four to six percent range that is usually used for real interest rates. The valuation of these costs and benefits is more problematic. Increased timber production is perhaps the easiest benefit to value. The important point is that the benefit of a management action is the increase in timber volume, quality, or both. On Ontario management units, a low level of silviculture will not eliminate timber production altogether. Even when natural regeneration is used, there will be some timber produced. The value associated with implementing artificial regeneration is seen in the increased and more rapid production of timber, and often in the regrowth of more valuable species.

Timber can be valued at an amount equal to Crown dues, although Crown dues are not closely linked to factors that affect timber value. A more appropriate approach is to use the net value created by processing the timber. This is calculated by taking the market value of the products obtained from a cubic metre of wood and subtracting the costs of harvest, transporting the wood to the mill, processing in the mill, and transporting the product to market. A study funded by MNR's Forest Values Group used this approach to value timber in Ontario (See SSK Inc., 1993)

In districts where tendered sales are common and representative of the local timber, the bid amounts provide a basis for estimating the value of standing timber.

The analyst must be careful to ensure that the BCA has the proper scope. Stand-level analysis is appropriate for landowners who are interested in single stand management and even for larger ownerships such as conservation authorities which do not require an annual timber harvest. Furthermore, stand level results can be usefully incorporated into forest level harvest scheduling programs when a relatively small proportion of the management unit area is affected by the treatment alternatives. However, if a manager is comparing treatments that will be applied to a significant number of hectares in a forest, then a forest level analysis is needed to properly account for between stand interactions. This type of analysis can be very complex and envolved and usually requires the use of computer modelling.

The intensification of a regeneration regime should yield a merchantable stand sooner than would otherwise be the case and presumably will contain a higher yield of a more valuable species. The benefits then will be the increased value of the future stand due to the higher stocking to the desired species and the earlier avialability of the areas for harvest.

Typically, the point at which a stand-level analysis begins is after the removal of the current stand. The rationale for this is that the money spent to regenerate the site (i.e. the investment) is related to objectives concerning the future stand, and it is on those merits that the regeneration expense should be evaluated. When spending more money on the harvest operation will reduce the costs of subsequent regeneration, the extra expense should be included as a regeneration cost, since this is the reason for it.

In a single stand analysis, one should ensure that there are no inconsistencies in the analysis caused by evaluating different alternatives over different time periods, as can happen when different actions will affect the stand rotation age. This can be avoided by examining a long series of repetitions of each alternative, effectively giving rise to an analysis period of infinite length. Alternatively, one could choose an analysis period based upon multiples of the harvest ages (e.g. compare two fifty-year harvest cycles under intensive management versus a one hundred year harvest cycle under extensive management). However, it is not always simple to find the required multiples, especially when a large number of options are being compared. This type of analysis is accomplished by considering the present net worth of a series of harvest values predicted infinetely into the future.

The value of wildlife is more difficult to assess. Forest management will likely shift the distribution of wildlife species in a stand or forest. Ideally, one would like to have estimates of how wildlife populations will be affected by alternative management actions. This is difficult, due to the number of species involved and the variations that populations exhibit over time. It is not yet possible to track all species-only those of importance or value to humans are tracked (although the criteria for importance are changing rapidly).

At the present, valuation of a major species poses significant obstacles. Suppose the moose population is locally important. Moose have value because people like to view them and because people like to hunt them. In theory, it is possible to survey people who live in and/or visit the area of interest, ask what value they place on having various population levels of moose, and adjust the results to account for the entire local (human) population and tourists. Statistical refinements based on income and level of camping or hunting experience are often made.

Studies in the U.S. have shown that a typical person will be willing to pay some amount for higher populations of many species because there are better wildlife viewing opportunities and because there is some satisfaction in knowing that there are more (rather than fewer) animals of that species alive in the area. A hunter will also be willing to pay some amount to have more game species in an area because the chances of a successful hunt are increased. A larger game population will also be likely to lead to an influx of hunters from other areas, which is a plus in a regional analysis but of little effect in a provincial or national analysis.

The author is not aware of any published data based on willingness-to-pay surveys that have estimated the value of wildlife species in Ontario. Where game species are involved, one alternative is to use the value of expenditures made by hunters as a proxy for the hunting value per animal. This has the obvious drawback of ignoring the value of benefits unrelated to hunting, but since expenditure data are available, they represent the best alternative for now. Of course, the benefit-cost analysis must still relate population level to degree of hunting effort and this will require some estimation.

The valuation of recreation opportunities poses similar problems. However, the key steps are the same as in the case of wildlife and timber. First, a quantitative relationship must be established between alternative management actions and recreation opportunities. Secondly, the value of a recreation opportunity, usually measured in terms of a recreation-day, must be estimated. One estimation technique is the willingness-to-pay survey. A more common technique, which is being employed by the MNR Forest Values Group, is known as the travel cost method. The central idea is to identify where the majority of recreation users come from, divide this region into zones of known distance to the recreation site, estimate the cost of travelling to the site from each zone, and then use the incidence of recreational use per zone and the zonal population to construct a demand curve for recreation at the site. Using site entrance fees is not appropriate for valuation purposes because prices are often set at low levels to encourage use. Finally, the valuation of such features as aesthetics is in its infancy, especially in Canada. Willingness-to-pay surveys of various types are again used in this case. Unfortunately, there are no known credible estimates of the value of aesthetic quality in Ontario.

There are two general classes of costs: out-of-pocket expenses and indirect losses or disbenefits. The out-of-pocket costs of undertaking various activities are usually unambiguous, but can be difficult to track down. Some judgement has to be exercised in the case of expenditures for fire protection and road construction and access. In these cases, a clear understanding of the reasons for making these expenditures is an important starting point. Fire protection is intended primarily to prevent the loss of human life, and secondly to prevent the destruction of buildings, machinery, or other capital goods. Therefore, only a fraction of fire protection costs can be charged against forest management.

At a forest level the choice of the exact fraction of the fire program's costs to allocate to forest management would depend upon the fire district, proximity to habitation and recreation values, wildlife concerns and of course the value of the forest and its contribution in providing wood or forest land supplies. Any attempt at a local division of these cost allocations would be artifical due to way these costs are accounted for at the aggregate program level, the interdependency of the social values associated with the use categories and the fact that in most areas fire prevention or control expensess are incurred to protect several use categories at one time.

Similarly, roads are built to facilitate timber extraction, subsequent assessment and regeneration, and provide access to the public for other purposes. Again, a subjective estimate of the proportion of the road value attributable to timber extraction is usually made. The costs of roads must also be amortised over some period of time, such as the 20 to 25 years often used for large capital expenditures such as pulp and paper mills.

As was mentioned above, costs of damage or reductions in non-timber values should also be included. These can be estimated using the principles that are appropriate for estimating non-timber benefits, described above.

The valuation of costs and benefits will depend on who is doing the analysis. A planner working for a Forest Management Agreement (FMA) holder will have a different perspective than a government planner on the benefits and costs associated with alternative management strategies. For example, since some regeneration costs are partially refunded by the government on FMAs, the company and the government will naturally place a different cost on regeneration. This is as it should be. A simple example will be presented which illustrates the use of benefit-cost analysis to compare the application of three forest-level strategies on a management unit. Suppose that these strategies are characterised as:

- Reliance on Natural Regeneration
- Extensive Management

- Intensive Management

The particulars of the strategies are unimportant; the key point is that each increase in management intensity results in a greater volume of wood being produced from the unit. Suppose that the wood supply levels are expressed as annual volumes. The table below shows, for each option, the annual wood supply, the value of the annual harvest, and the annual cost of management. Note that wood has been valued at $10/m3, and non-timber impacts have been excluded.

It will be assumed that some harvesting will be done, and so one of the three options must be selected. The key issue is the comparison of the net gains from management intensification. The table below shows the PNW and BCR associated with each option.

A comparison of PNW's shows that the greatest net benefit is associated with intensive management. Therefore, if sufficient funding is available, intensive management would be preferred. However, the BCR declines as management is intensified from extensive to intensive levels. This indicates that the gain per dollar spent decreases as the management level is increased.

If there are other potential investments that yield a BCR ranging from 1.46 to 1.60, then the greatest net benefit will be obtained by adopting extensive management and investing the remaining available funds in these other projects.

In this way, the BCR acts to guide investment so that the greatest overall PNW can be obtained. In this example, the profitability of management was at least partially justified by the increase of current harvest made in the expectation of future yield increases due to management. This is the allowable cut effect (ACE). Economists observe that the ACE has often been used to justify poortreatment investments (see Davis and Johnson, 1987, Pearse, 1990). It is considered that there are really two separate decisions involved-the selection of the harvest level and the selection of the appropriate silvicultural treatment. Mainstream economists argue that these decisions should be analyzed separately on their own merits, whereas forest economists point to sustainable yield requirements as the basis for linking the two decisions (Hyde, 1980).

In Ontario, the argument is clouded further by the public ownership of forests and the flow of non-timber benefits that come from them. This implies that the government should include in the analysis the social benefits that can be expected to arise from an increased harvest level.

What conclusions can be drawn from this debate? The first is that one should be reluctant to include the ACE when analyzing projects on private land where the sole purpose of investing is to grow timber but ACE is a valid factor to include in analyses of public forest management.

The second point is to heed the economists' warnings and critically examine the merits of the silvicultural projects being considered. Foresters should be careful about planting on medium and poor site classes, and on sites distant from the mill. Due to the additional costs of hauling it to the mill, the value of timber grown further from the mill is lower than the value of equivalent timber close to the mill.

One of the great difficulties in forest planning is dealing with the long time periods required to grow timber. The fastest growing boreal stands may be harvestable in 40 to 50 years and it is difficult to forecast real timber prices that far in the future with any accuracy. The uncertainty inherent in predictions for 100 years or more is enormous. Therefore, in view of the significant degree of uncertainty associated with assessing the profitability of a regeneration project, one may question whether such an exercise is worthwhile.

Nevertheless, most people believe that planning is useful and planners can only use their best estimates. It is also the case that accurate forecasts of future prices and costs are less important when one is using BCA to select the most attractive project or to allocate funds among projects.

In these situations, any errors in forecasting will apply equally to similar projects and reservations about long-term forecasting are less relevant.

Forestry debates present society with several choices that need evaluation. The task is to estimate the value in each of the choices presented. The purpose of this paper is to establish a framework that will allow each environmental policy option to be systematically and comprehensively assessed for its impact on the total economic welfare of the community.

Such a framework consists of ensuring that the non financial values, financial values and alternative financial values derived from the forest are accurately evaluated, and hence given their full and due consideration.

There are presently significant barriers to the achievement of this objective. The three barriers identified in this report are:

1. The existing inability to determine the correct financial values for forestry operations.
2. An inability to determine the correct financial worth of alternative activities such as plantations and water production.
3. The lack of a methodology which determines the relative worth of non financial values (NFV's) of the forest resource.

Removal of these impediments, which is an exercise in microeconomic reform, will bring improvements in the financial efficiency of the Australian economy, whilst improving the overall quality of life for Australians in general. The first two impediments can be removed by the utilisation of some fairly common methods. With specific reference to the third impediment, however, there are many approaches by which appraisal of non financial values can occur.

Some of these are specific methodologies such as contingent valuation and public ranking of options. Other methods centre around public involvement in the assessment of options. Yet other methods are based on using 'rules of thumb' to account for non financial values.

Among the reports surveyed, some techniques are mentioned quite frequently. These included public participation as a key necessity for getting non financial values properly assessed. Surveying of public opinion, systematic reviews of natural features and the use of experts also seemed to be relatively common.

The task therefore becomes one of finding an appropriate route through the range of techniques offered.

Section 3 of this report briefly outlines the relationship between the concepts of externalities, economic value and community well being. The absolute requirement of good economic analysis is the need to accurately appraise all values, including externalities (non financial values) in any assessment framework. Economics provides the theoretical standard by which the existing assessment frameworks can be judged.

Section 4 reviews several major forest reports in order to identify four significant components of forest project appraisal. These include the values of the forest, the valuation methodologies proposed for assessing values, the pricing/ financial valuation of forestry and alternative activities, and the assessment framework for appraising policy options in the search for the socially best outcome.

In particular, Section 4 examines the role of pricing in the appraisal of forest policy options. The existing assessment processes and the various forest reports (apparently on the grounds of practicality) adopt a 'compartmentalised' approach to pricing. In other words, protection of non financial values is carried out prior to, and separate from, the pricing of forest timber. If the appraisal process is incomplete, then the pricing of timber is inadequate by definition. Pricing policy is significant for estimating value to the extent that it partly determines the financial value of forest timber.

Section 5 then describes weaknesses in the existing appraisal processes. These flaws range from a broad neglect of non financial values in project assessment to more specific issues. These include the failure to comprehensively identify and value non financial values, often even in the form of physical impacts; failure to identify or consider alternative financial values and alternative project options; and the failure to provide accurate statements about the financial viability of projects.

This report makes clear that there is indeed evidence of serious weaknesses in the determination of the financial values of native forest timber. Our analysis indicates the very strong likelihood that native forest timber harvesting, as a commercial activity, is being subsidised (eg. low cost public sector loans, no commercial charges for land), is inefficient (eg.. there are better returns to be had from harvesting water in at least one case) and poorly managed (eg. the agencies are unable to provide appropriate accounting information on their commercial activities).

Consequently, a significant microeconomic reform process needs to be undertaken, if the correct financial values for native timber sales are to be determined.

Inappropriate appraisal of financial values may also be suppressing economic growth in the private forestry and agroforestry sectors (and other alternative/substitute financial values). The lack of a 'level playing field', with respect to the cost of business inputs (eg. land and capital) between the public and private forest managers, suppresses investment in private forestry at a lower level than that which could normally be expected.

If financial values and alternative financial values of the forest were to be appropriately appraised, this may identify alternative revenue streams and investment opportunities for hard pressed forestry commissions.

Summarised, inaccurate financial values and a limited knowledge of alternative financial values combine to undermine the basis of comparison between the financial and non financial values of the forest. This problem is compounded by the lack of a methodology for determining the relative worth of non financial values.

In conclusion, it must be reiterated that any assessment of the social benefits and costs in forest policy must include three pieces of vital data: an accurate appraisal of financial values, alternative financial values and non financial values. Without reference to such data, environmental decisions become less an attribute of good management, than of reliance on sheer blind luck.

Resolving these problems will contribute to better environmental management, greater financial performance and more jobs. This makes reform in this area a key part of the process of microeconomic change, resulting in better environmental and economic policy.

VALUATION OF THE MANGROVE ECOSYSTEM

Due to shortage of time and fund, the project focuses on partial analysis, including cost-benefit analysis of planting, protection, thinning and extraction options, fishing and part of benefits rising from eco-tourism, while indirect-values or function and existence values will be initiate for further studies.

All marketable data available has been collected and collated, and the solicited by the intensive surveys in December 1999 and January 2000. A set of secondary data available in Management Boards of Can Gio Forestry Park, Statistic Office of District Can Gio and Management Boards of Protected Forests is being accessed, analyzed and aggregated by computerization. A new approach on economic valuation based on Total Economic Value (TEV) is introduced in the case study. The analysis framework of direct use values and non-use values of mangroves which is developed by using Stella 2 dealing with ecosystem services of mangroves in can Gio is tested. A detailed description of methods estimating the direct value of mangroves is described, including their limitation.

Based on Cost-Benefit Analysis, it is presented for each direct benefits/costs: planting, thinning, branching, nipa thatching, propagule supply, in-site and off-site fishing catch, aquaculture, fishing trades and travel cost analysis of eco-tourism. Some avoided costs / benefits is not estimated for shoreline protection, flood and storm mitigation and stock of water. However, the value of biodiversity conservation is evaluated by using the cost or annual investment from HCMC, as considered the existence value. It needs to be added and improved.

The results of cost benefit analysis (CBA) shows that it is of a partial nature, comparing planting, protection and thinning costs, with the direct benefits from extracted marketable products, and with the indirect benefits of the City' investment to restoration of the biosphere reserve. A benefit to cost (B/C) ratio in range of discount rates from 1 to 10 % is analyzed. For the B/C Ratio of direct benefits, the figures illustrates that most of the costs for planting from initial stage and protection and management system is less than the direct benefits. It can be explained that the limitation of thinning areas, over exploitation of fishing catch, water pollution, oil spill and others related is contributing the low direct benefits, but still higher than cost. However, this is also not meaning that local livelihood of people can be able to sustain in terms of basing the direct benefits.

Otherwise, a new regulation of banning any thinning activities from the City Administration has been launched in September last year and enforcement from early this year. According the regulation, a triple incentive will be given to households allocating forestry lands and protecting forests. However, a detailed response from communities has not been recorded. For the B/C Ratio of total benefits, the big different of benefits to costs is showing the vital role of investments from the National Government and City Administration to restore the reserve.

Based on the estimation of total economic value of mangrove restoration and rehabilitation, management options in dicision-making processes can be laid out from current to hypothetical scenarios in order to create a scene from unsustainable development up to sustainable one for resource use, like mangroves. Humain activities as driving forces effect to both socio-economic and natural processes.

The total economic value can be used as a tool in estimating the activities, especially the mass conversion of mangroves to other purposes. It was really an exchange of three kind of benefits (local, regional and global) in sustainable to one private benefit in short time. As an economic indicator, the total economic value may reflect the effects of policies in management and development in feedback estimation.

This is initial work to contribute the analysis framework needing an investment of time and funding, especially indirect use value, option and quasi-option values and existence value or non-use value as described by Barbier (1989, 1993, 1994) and Scodari (1990). The estimation of non-market values is a big challenge of environmental economists. However, valuation methods proposed by R. Costanza et al. 1997 on ecosystem services, natural capital and GNP are being applicable for this issue. It would be so kind to ask UNESCO/ MAB to invest more time and funding for the work.

The Can Gio Mangrove Biosphere Reserve has been designated for inclusion in the World Network of biosphere Reserve, by decision of the International Coordinating Council of the Program on Man and the Biosphere, in January 21, 2000.

Consisting of 77 mangrove species (35 true mangroves and 42 associates), the mangrove has contributed an economic and environmental importance in the coastal zone, as long as very vulnerable to impacts from natural and human driving forces. For almost population living in these areas, the mangrove provides not only valuable products from forests including timber, firewood, charcoal, tannin, food, medicinal but also the breeding ground for many species of marine organisms like shrimp, crab, fish, water and migratory birds as well as other economically terrestrial species of monkey, wild boars, boas etc.

It has been recognised by economists that the functions and services provided by mangroves, and wetlands in general, have positive economic value and that these are often ignored in the ongoing process of mangrove conversion (Barbier, 1993; Ruitenbeek, 1994; Turner, 1991; Swallow, 1994; Farber and Costanza, 1989).

The various functions and services provided by mangrove areas have been documented and appraised (e.g. Mitsch and Gosselink, 1994; Reimold, 1994; Lugo and Snedaker, 1974). Mangrove wetlands display the features of public goods in that their use is non-exclusive, and they are converted to other uses with private economic returns because the functions and services associated with them are undervalued. Identification of the functions and services and the incorporation of these into policy and the designation of property rights are therefore necessary first steps in promoting sustainable utilisation of such resources. The valuation of functions and services has often been expressed in a framework of the total economic value.

Critical issues in promoting the adoption of such schemes, and hence their ultimate sustainability, include the timing of the costs and benefits and the sensitivity of the economic appraisal of such schemes to discount rates and the limit of the benefits of rehabilitation compared to conservation of existing wetlands (reflected in the absence of option and existence values for such habitats). In the instance under analysis here, the benefits of direct use value of timber, non-timber products including fish, shrimp, crab and thatching, and other benefit for recreation, biosphere reserve as well is an important issue

THE STUDY SITE

The Biosphere Reserve is located on Can Gio District, about 65 km south of Ho Chi Minh

City with latitude: 10° 22'14"-10°

40'09" and longitude: 106°

46'12"-107°

00'59". It is 35 km

long from North to South and 30 km from East to West. During the two Indochina wars, almost all the mangroves in Can Gio were destroyed. After many years of herbicide spraying, the degraded land here still remains shattered bushy or bare land. A great effort of local people after wars is the rehabilitation of 21,000 ha of mangroves.

Can Gio is situated in the typical tropical monsoonal zone with 2 distinctive seasons. The dry season lasts from November to the next May, the rainy season from the end of May to the end of October. The temperature is high, stable, the monthly mean temperature is 25.5-29.0° C. The daily average temperature amplitude is 5-70C. The insolation is 5-9 sunshine hours per day. The solar radiation is 10-14 kcal/cm2/month on an average. The radiation intensity does not vary much with different seasons.

The monsoons form different wave directions and coastal currents, directly influencing Can Gio's coast, leading to erosion and sea level rise. Compared to other areas in Ho Chi Minh City, the rainfall in Can Gio is lower and decreases gradually southwards. The rainy season begins on 20-25 May and ends around 25-31 October. The average annual rainfall is about 1,336mm. September sees the highest rainfall (300-400mm). The mean annual relative humidity is 80%. In the dry season, the air humidity during daytime is usually under 60%. The evaporation is 3.5-6mm/day and in March and April 7-8mm/day at the highest. It is 2.5-5.0 mm/day in the rainy months

Salinity: The survey data after 11 years (1977-1988) (Thanh, 1988) showed that in the dry season, highest salinity in March, April being 1.9-2.0% at Nha Be area, 2.6-3.0% near the sea. In the rainy season, the salinity is low; from August to October, the salinity in mangrove areas is only 0.4-0.8%. The average monthly salinity is 1.8%. Since 1988, the drainage of water from the reservoir of Tri An Hydroelectric Plant in the dry season has decreased the salinity in canals and rivulets of this area.

Given the prevailing circumstances in the coastal districts of Can Gio, and similar regions elsewhere, it is clear that mangrove rehabilitation can have a variety of benefits. In such situations, mangrove rehabilitation can provide income where households are often severely constrained in cash income sources, as well as bringing about environmental benefits in terms of productive assets and biodiversity resources. The following sections quantify an economic model of such an approach.

METHODOLOGY/APPROACH

The approach of total economic valuation is used in this work. Total Economic Value (TEV) is given by the sum of a number of components:

TEV = Direct-use value + Indirect-use value + Option value + Existence value

Direct use values include revenues from timber and values of other products such as thinning from Nipa palms. Indirect-use values or 'functional' values relate to the ecological functions performed by mangroves, such as global geochemical cycling, the protection of agricultural areas, and the provision of spawning grounds for fisheries. Option value, or more precisely, quasi-option value is the expected value of the information on the benefits of an asset, conditional on its preservation enabling an increase in the stock of knowledge relevant to the utilisation of the asset. A frequently evoked example of quasi-option value is associated with genetic resources: for example, the future value of pharmaceuticals developed from plant materials.

Existence value relates to the value of environmental assets irrespective of current or optional uses. Empirical measures of existence values based on donations to conservation organisations, or on the contingent valuation method, suggest that these can be a significant element in total economic value, especially in contexts where the asset has unique characteristics or cultural significance.

Although intuitively appealing, applications of Total Economic Value to particular environmental resources are few because of the problems of double counting if the elements are simply added and of quantification of the components. For example, although direct and indirect benefits of mangrove conservation can generally be assessed, the option and existence value of mangroves are difficult to determine within an economic analysis. Much research has been undertaken in this area, and some studies suggest that non-use values (option and existence) can be at least as large in magnitude as use values (see Walsh et al., 1984; Cummings and Harrison, 1995 for example). It has however been pointed out in studies of Total Economic Value that option and existence values reflect global as well as local scarcity and that the welfare gained from resource conservation normally accrues at the regional or global level, bypassing the local users (see Adger et al., 1995, and Kumari, 1995, for example).

The crucial aspects of value for local decision-making, and for the differential impacts of global change, are the direct and indirect use benefits. It should be noted that some economic benefits of the mangrove resource will increase in value over time, while others will remain constant or decline. For example, as agricultural development intensifies, the potential economic losses from storm surges increases, so the value of the coastal protection function of the mangroves will rise accordingly. Exogenous environmental change associated with global climate change may increase the frequency and intensity of storm surges, and hence the value of this function of the mangroves will rise.

Recent attempts to value coastal wetlands have concentrated on valuing specific contribution of various wetland systems.Some economics studies have valued the benefits of temperate wetlands, but to date, little analysis of tropical wetland benefit has been undertaken (Turner et al.1991,92). The value of Florida coastal marshes for marine fishery production, relating wetland area to crab production, in this case the Florida blue crab fisheries was estimated by

Lynne et al. (1981) and the value of Louisiana wetlands for hurricane wind damage protection was estimated by Gosselink et al. (1974).

E.B. Barbier (1993) gave a basic methodology for assessing and valuing the economic benefit of tropical wetlands applied in Central America in distinguishing between direct-values, indirect-values and non-use values (typology of benefits of conservation), particularly a case study of a mangrove system in Indonesia for woodchip production. The value of wetlands for residential development and the value of Virginia coastal wetlands for oyster production. The valuation of oyster production is directly related to wetland area. The same methodology can therefore be used to value for crab and shrimp fisheries, but not for offshore fisheries were estimated by Bertelson and Shabman (1979).

S. Farber and R. Costanza (1987) stated that the harvest of the shrimp and crabs from the mangroves is directly related to the area of the mangroves by relating the marginal value to the area in a regression equation. The paper derives a linear relationship between area of wetland and offshore fish catch based on a cross section of wetlands. Ruitenbeek, H.J. (1994) gave a full valuation excise, and show the trade-off between protection and logging mangrove forest for timber in indicating that harvesting is much lower when the 'linkages' between the reduction in mangrove area and the functions (such as off-shore fishery) are taken into account. The benefits of the mangroves remaining in existence include the shrimp fishery and biodiversity (Batie and wilson, 1978).

Natural wetlands perform many important functions for humankind-storm prevention, flood and water flow control, and wood products. When properly measured, the TEV may exceed the economic gain of converting the area to an alternative use. It is difficult to assemble estimates of values of these various contributions for the same wetlands system.

Because of this difficulty, little is known about the ordes of magnitude of value for one entire wetlands system. Economic analysis of resource use can be undertaken in order to assess the magnitude of benefits to local users of the resource. Some values of the goods and services can be assessed by observation of existing markets, but some of the functions and services of mangroves are indirect, or functional, benefits (see Pearce and Turner, 1990, for example) is still difficult to estimate. The purpose of this paper is to obtain a consistent set of estimates of the value of one wetland system, a biosphere reserve for as many independent contributions as existing methodology and data currently allow.

FRAMEWORK FOR ANALYSIS AT THE STUDY SITE

The valuation of the mangroves is divided into two components: direct use values and indirect or non-use values. For each component, potential counts of benefits as well as their linkage, where the STELLA 2 is basically used for potentially linking components and estimating afterward. For direct use values, all marketable estimations is laid out as thinning, branching, nipa thatching and propagule supply contributing to local benefit 1 though the cost benefit analysis 1. The local benefit 2 includes on site and off site catches, aquaculture at household levels and trade activities of the aquatic products. Clearly, these benefits is linked to other bio-ecological processes such as recovery and resilience of mangroves through biomass and productivity, and nursery and refuse habitats on which the fisheries depend. All linkages should be estimated by ECOPATH application.

For indirect or non use values, the regional benefit that includes air clean, ground water and domestic water supply, and shoreline protection is linked with ecological functions such as gas regulation, recycling of water and erosion and flood control. The global benefits of biodiversity or existence values includes eco-tourism and knowledge benefits. It is a part of eco-tourism may be monetary estimated but the rest and their cultural and information values should be included in the existence values.

Obviously, the plan would be useful to have a analytical framework for study and contribution to management and decision making processes, but we can not evaluate all components presented in the diagram, so a partly estimation is our expectation for doing the project. The linkage should be done for next phases.

PARTICIPATORY ECONOMIC EVALUATION

Context: In the transitional period from the planned to the market economy, the Government of Vietnam has encouraged re-allocation of barren land for smallholder forest resource development. Under the framework of the recent 5-Million-ha Reforestation Program supported by diverse donor agencies, the German-based Kreditanstalt fuer Wiederaufbau (KfW) is co-financing several forestry projects to support smallholder establishment of production forests with protective functions on barren land. Projects are executed by the working group of two consulting firms (GFA Terra systems and GWB).

Forest ecosystem services in the project area are complex in nature and highly site-specific. This makes it difficult to generalize benefits across sites. Also local decision makers and villagers are sceptical about results of economic analysis that is conducted in a style which is not transparent enough for people to understand. People simply do not trust general figures. For instance, they distrust the idea that the environmental value of their forests could be so great in monetary terms. The limitations of conventional valuation suggest that it might be a good thing to encourage active involvement of villagers in the economic valuation of their forests. However, this is a new field in participatory approaches in land use development. To date little practical experience in this field exists in developing countries.

Participatory Economic Valuation (PEV) procedure: The purpose of PEV by the project is to assist forest smallholders and local leaders to improve forest resource development decisions. The basic assumption of PEV is that the best currently available information on the costs and benefits of different land uses is held by local decision-makers, technicians and land users.

The advantages of PEV-compared to conventional valuation approaches-are: (i) PEV fully utilizes the data and knowledge of local decision-makers and villagers; (ii) PEV maximizes the involvement and feedback of these groups; (iii) PEV raises awareness about the importance of forest valuation and knowledge of how to do it. To enable local people to fully participate in valuation, PEV must be conducted in a style which (i) is simple and transparent enough to be understood; (ii) is based on data perceived as accurate; and (iii) uses assumptions perceived to be reasonable.

This can be achieved by building on local knowledge (e.g. of perceived physical environmental impacts in terms of change in yield of rice production by villagers) and economic data and assumptions, combined with field checks with villagers. The core focus of PEV is village valuation by forest extensionists with a group of villagers and commune (township) representatives. All steps in the calculation and results must be visualized and reconfirmed by the participants in the village valuation. It is important to include women in the valuation exercise to ensure that their perspective is not missed. Uncertainties about assumptions made in calculation are addressed by calculating two scenarios (high and low value) for each forest model.

PEV tools used in the PEV exercise included the 3-D model (a three dimensional image of the village), topographical maps, land use maps, ruler, pens, blackboard, and large sheets of paper (A 0).

Forests provide environmental benefits from the village to global scale. The scope of assessment was limited to forest environmental services (FES) for which future local payment is not unlikely. For that reason environmental forest services from which the global community will benefit, but payment is currently not feasible (like carbon sequestration), have been excluded in valuation.

The PEV procedure consists of a minimum of 7 steps, beginning with preparation of background data like collection of secondary yield and market data (step 1), followed by briefing villagers about the purpose, scope and expected results of PEV and collection of basic village data including forest resources (step 2). Then basic village data are collected (step 3). To get a first overview of what type of forest values the villagers perceive, the major direct benefits, forest environmental services, optional values, and non-use values in the village and beyond are identified (step 4). It is important to discuss whether men and women have different perceptions of forest values. To reconfirm the identified forest values, sites are visited and the physical impact area is calculated (step 5). Then, the direct benefits and costs of predominant forest models in the village are calculated, subsequently also environmental services, optional values and non-use values are valued (step 6). Finally the results of the PEV exercise are discussed at a village meeting and recommendations for further actions are made by the villagers (step 7).

PEV field exercise: In December 2001 the PEV concept elaborated by the author during a KfW short term assignment was refined with key project staff and then tested with project extensionists in seven villages in the Provinces of Bac

Giang and Lang Son in North Vietnam. For testing the PEV approach only villages were selected where a significant area of the forest canopy was already closed. Results of Exercise: Forest values identified were direct benefits (timber and non-timber products like star anise and resin), environmental benefits (protection of irrigation facilities, improvement of farm land and fish productivity), optional values (forest land reserved for future land use mainly for tree crop development). "Spirit forests" represent the only forest value independent of their actual current and future use (non-use value). Reducing soil erosion and settling sediments in small water reservoirs are the most visible impacts of afforestation of barren slopes. The rate of exchange used was 1 EUR equivalent to 13,000 VND.

Results of calculation of forest environmental services (FES) included: o Enhanced paddy productivity (15,600 to 600,000 VND/ha/year or 1.2 to 46 EUR/ha/year). The highest value is equivalent to the farm gate price of a big pig or the market value of paddy (unshelled rice) from 0.1 ha. o Annual reduced costs for extraction of sedimentation from micro-irrigation facilities (225,000 VND/ha or 17.3 EUR/ha); and Increased annual fish productivity in small village ponds (41,900 VND/ha or 3.2 EUR/ha). This is equivalent to the farm gate price of 4 kg fish.

Optional values for forest land reserved for tree crop cultivation ranged between 8.0 and 94.5 million VND/ha (615 and 7,270 EUR/ha). "Spirit Forests" (non-use value) can have a high value in terms of biodiversity, but villagers did not dare to give a monetary value. The undiscounted Total Economic Value (TEV) per village household in one sample village was between 23.2 and 39.9 million VND/household (1,785 to 3,070 EUR/household) and the discounted TEV between 5.3 and 11.2 million VND/household (410 and 860 EUR/household).

For the high value scenario the undiscounted FEV was around 6% and optional value 1%. The FES for one ha of pine plantation discounted was 14% and FES undiscounted equivalent to 6%. One example of Willingness to Pay (WTP) for disproportionately perceived private environmental forest services (i.e. increased fish pond productivity due to improved water supply and quality from forests) was 10% of the incremental income due to the FES.

Perception of environmental forest benefits was highest where villagers had experienced a period with natural forests, followed by deforestation and unsustainable upland land use including subsequent reforestation. Especially elder villagers highly appreciated the environmental benefits of forests. From the viewpoint of villagers, the distinction between production and protection forests as defined by Vietnamese Law does not make sense in a situation where a production forest in a village also has a significant protective benefit (e.g. when a forested micro-watershed directly protects and improves the productivity of a significant area of paddy land).

Lessons Learned: The exercise in Vietnam reconfirmed that the best realistic economic data for existing forest models can be gained from technicians and farmers, especially those who have developed local forest practices.

The tested bottom-up approach of valuation also proved to be a good medium of communication to identify the production forests in the village for which clear cutting should be avoided at all costs in order to maintain the watershed benefits. In this case PEV provides economic arguments for conversion of unstable pure plantation to continuous forest cover.

6

FOREST BASED INDUSTRIAL DEVELOPMENT

INTRODUCTION

There is no industrial forestry activity apart from small scale charcoal production (largely for tourist and domestic recreation pursuits), extraction of wattle for fish and lobster traps, and some use of mahogany and white cedar for boat construction and fence posts. Estimates are that 1135 tonnes/year of charcoal and 4000 fence posts/year are produced, but there are no supporting data. The wood related manufacturing sector is also small, but again there are no records. Some opportunity may exist for handicraft material for carving, but technical training would be required to enhance growth in this area, particularly to utilize the limbs pruned from urban mahogany trees, some of which are dying or so badly mutilated that they should be felled and replaced. In addition, there is potential for non-wood products of agroforestry such as fruits and nuts, fodder and water.

Neem has been quite successful in Antigua and Barbuda and is widespread, mainly along road-sides or fence lines. It seems to have withstood the effects of hurricane Hugo in 1989, and attention should be given to determining the utilization prospects for this valuable fast-growing species. It is being planted elsewhere in the world and produces a wide range of products.

These include the extract from the seed kernel and seed coat to yield neem oil, which is used as a lubricant, insect repellant and insecticide and in the manufacture of soaps and cosmetics. The seedcake residue is a valuable organic manure and cattle feed, while the root exudate has medicinal properties; the bark yields an industrial gum and the leaves are used as a green manure. The timber is also useful for lumber and fuelwood. Antigua and Barbuda has an excellent opportunity to investigate the industrial potential of this tree which could attract investment to the country.

"Liberty and forest laws are incompatible," remarked an English country vicar, speaking on behalf of villagers shut out of woodland reserved for the exclusive use of the king, in 1720. The history of state forestry is indeed a history of social conflict.

In monarchies and in democracies, in metropolitan Europe as well as in colonial South Asia, the state management of forests has met bitter and continuous opposition. On the one side are the professional foresters who believe that timber production can be ensured only through the exclusion of humans and their animals from wooded areas; on the other, the peasants, pastoralists, charcoal ironmakers, basketweavers, and other such groups for whom access to forests and forest resources is crucial to economic survival. Environmentalists have added to the criticisms of these latter groups, charging foresters with simplifying complex ecosytems in the direction of commercially valuable but biologically impoverished monocultures.

These contending parties have battled for more than two hundred years. In continental Europe, the eighteenth and nineteenth centuries were peppered with social protest movements against the state management of forests. These protests inspired, among other things, Karl Marx's first political writings and a memorable novel by Honore de Balzac capturing peasant hostility to forest officials. When the European model of strict state control over forests was exported to the colonies, the disaffected peasants and tribals responded with arson and violence. Movements over forest rights were a recurring phenomena in colonies ruled by the British, the Dutch, and the French. The conflicts persisted when the post-colonial governments of countries such as Malaysia and Indonesia followed the authoritarian model of forest management inherited from the colonizer.

In recent decades, however, the global discourse on forestry has moved towards a more accommodationist perspective. Foresters and peasant protesters now seem to talk to, rather than talk past, each other. A willingness to listen to and at least partially incorporate the other point of view has replaced the rigid and uncompromising attitude of the past.

Within the forestry profession itself, skeptics doubt the contemporary relevance of the custodial and policing approaches previously followed. A system of natural resource management crafted in absolutist and colonialist times clearly needed to be seriously modified or even overthrown.4 Social activists and community leaders have, meanwhile, moved from demanding a total state withdrawal from forest areas to asking governments to more seriously and sympathetically consider the rights of forest-dependent communities.

With this move from conflict to collaboration have come shifts in the language of forestry itself. Terms such as "scientific forestry" and "rational land management," euphemisms for state control and commercial timber production, are being rapidly replaced by sweet-sounding phrases such as "community management," "participatory development," and "joint forest management." While these terms have come into vogue in the last two decades, they have, in fact, a very long geneaology.

From the beginnings of state forestry, there have been serious attempts to democratize the regimes of resource management. Both dissidents within the bureaucracy as well as intellectual activists outside it tried hard to make the state respond more sensitively to the just claims of local communities. The ongoing programs of joint forest management in India can draw legitimacy and sustenance from a struggle that is at least a century old.

THE LAW AND THE PROTESTS

The crucial watershed in the history of Indian forestry is undoubtedly the building of the railway network. In a famous minute of 1853, the governor general of India, Lord Dalhousie, wrote of how railway construction was both the means for creating a market for British goods and the outlet for British capital seeking profitable avenues for investment. Thus between 1853 and 1910 more than eighty thousand kilometers of track were laid in the subcontinent.6 The early years of railway expansion witnessed a savage assault on the forests of India. Great chunks of forest were destroyed to meet the demand for railway sleepers (over a million of which were required annually). The sal forests of Garhwal and Kumaun, for example, were "felled in even to desolation." "Thousands of trees were felled which were never removed, nor was their removal possible."

This depradation brought home most forcefully the fact that India's forests were not inexhaustible. At this time the British were unquestionably the world leaders in deforestation, having burnt or felled hundreds of thousands of acres of woodland in Australia, southern Africa, northeastern United States, Burma, and India. Knowing little of methods of sustained-yield forestry, they called in the Germans, who did. Thus in 1864 they established the Indian Forest Department, which for the first twenty-five years of its existence was serenely guided by three German inspectors general of forest-Dietrich Brandis, Wilhelm Schlich, and Bertold von Ribbentrop.

For its effective functioning, the new department required a progressive curtailment of the previously untrammeled rights of use exercised by rural communities all over South Asia. An act was hurriedly drafted to establish the claims of the state to the forestland it immediately required, subject to the provision that existing rights not be abridged. This act was "infinitely milder and less stringent than that which is in force in most European countries." The search commenced for a more stringent and inclusive piece of legislation. In 1869, the Government of India circulated to the provinces a new draft act, which sought to strengthen the state's control over forest areas through the regulation and in some cases extinction of customary rights.

The new legislation was based on the assumption that all land not actually under cultivation belonged to the state. Of course, it was not easy to wish away the access to forests exercised in centuries past by peasants and other rural groups. The colonial state, however, argued that such use, however widespread and enduring, had been exercised only at the mercy of the monarch. Unless it had been expressly recorded in writing, customary use was deemed to be a "privilege," not a "right." And since the British government was the successor to Indian rulers, the ownership of forests and waste was now vested in it. "The right of conquest is the strongest of all rights," emphatically remarked one forest official. "It is a right against which there is no appeal."

There were, however, some notable dissenting voices within the colonial government. Sent the draft forest bill by the Government of India, the Madras Government in turn invited responses from various officers. The views of Narain Row, Deputy Collector of Nellore, are representative. The proposed legislation, he said, had no historical precedent, for "there were originally no Government forests in this country. Forests have always been of natural growth here; and so they have been enjoyed by the people." Another Deputy Collector, Venkatachellum Puntulu, of Bellary, argued that the burden of the new legislation would fall most heavily on the poor. While large landlords would find it relatively easy to deny the state any claim over their forest property, unlettered peasants would not be able to prove rights of ownership, even though they traditionally used forests as common property.

Criticizing the detailed rules prohibiting the collection of different kinds of forest produce, Puntulu penetratingly remarked that "the provisions of this bill infringe the rights of poor people who live by daily labor (cutting wood, catching fish and eggs of birds) and whose feelings cannot be known to those whose opinions will be required on this bill and who cannot assert their claims, like [the] influential class, who can assert their claims in all ways open to them and spread agitation in the newspapers."

After several such responses came in, the Madras Board of Revenue told the Government of India that the claim of the state to uncultivated forests and wastes was virtually nonexistent:

There is scarcely a forest in the whole of the Presidency of Madras which is not within the limits of some village and there is not one in which so far as the Board can ascertain, the state asserted any rights of property unless royalties in teak, sandalwood, cardamoms and the like can be considered as such, until very recently. All of them, without exception are subject to tribal or communal rights which have existed from time immemorial and which are as difficult to define as they are necessary to the rural population.... [In Madras] the forests are, and always have been common property, no restriction except that of taxes, like the Moturpha [tax on tools] and Pulari [grazing tax] was ever imposed on the people till the Forest Department was created, and such taxes no more indicate that the forest belongs to the state than the collection of assessment shows that the private holdings in Malabar, Canara and the Ryotwari districts belong to it. The Madras Government advanced three basic reasons for rejecting the bill drafted by the Government of India:

First, because its principles, scope and purpose are inconsistent with the existing facts of forest property and its history. Second, because, even if the Bill were consistent with facts, its provisions are too arbitrary, setting the laws of property at open defiance, and leaving the determination of forest rights to a Department which, in this Presidency at all events, has always shown itself eager to destroy all forest rights but those of Government. Third, because a Forest Bill, which aims at the regulation of local usages ought to be framed, discussed and passed by the local legislature.14

The objections were disregarded, and in 1878, the new bill passed. The act divided the forests of the subcontinent into three broad classes. State or reserved forests were to be carefully chosen, in large and compact areas that could lend themselves to commercial exploitation.

The constitution of these reserves was to be preceded by a legal settlement that either extinguished customary rights of user, transferred them as "privileges" to be exercised elsewhere, or, in exceptional cases, allowed their limited exercise. In the second class, of "protected" forests, rights and privileges were recorded but not settled. However, all valuable tree species were to be declared as "reserved" by the state, while the Forest Department had the power to prohibit grazing and other ostensibly damaging practices.

The Forest Act also provided for a third class of forests-village forests. But as these lands had first to be constituted as reserved forests, the procedure aroused suspicion among the villagers, and this chapter remained a "dead letter."15 Meanwhile, the area of forests under strict state control steadily expanded. In 1878, there were 14,000 square miles of state forest. By 1890, this had increased to 76,000 square miles, three-fourths of which were reserved forests. Ten years later, there were 81,400 square miles of reserved forests and 8,300 of protected forests. Given increasing demand for wood products, the state sought to establish firmer control over forests, both by expanding the area taken over under the Forest Act and by converting protected forests to reserved forest. The Indian Forest Act of 1878 was a comprehensive piece of legislation that came to serve as a model for other British colonies.16 Within India, it allowed the state to expand the commercial exploitation of the forest while putting curbs on local use for subsistence. This denial of village forest rights provoked countrywide protest.

The history of colonial rule is punctuated by major rebellions against colonial forestry-in Chotanagpur in 1893, in Bastar in 1910, in Gudem-Rampa in 1879-80 and again in 1922-23, in Midnapur in 1920, and in Adilabad in 1940. These rebellions sometimes extended over several hundred square miles of territory, involved thousands of villagers, and had to be put down by armed force. Even where discontent did not manifest itself in open rebellion, it was expressed through arson, noncompliance, and breaches of the forest law.

The participants in these protests were unlettered peasants and tribals, and we know far more of their deeds than their words. Nonetheless, their voices do figure here and there in the archives of the state, sometimes mediated by the language of the officials reporting them. Thus in the

1880s, when the government of the Bombay Presidency was aggressively demarcating the rich teak forests of the Dang district, preparatory to their constitution as state reserves, a Bhil tribal chief sent in a petition stating that "we do not wish to let the Dang jungle [be] demarcated, for thereby we shall lose our rights and we and our poor rayat [cultivators] shall always be under the control of the Forest Department and the Department will always oppress us." Around the same time, the colonial state was attempting to take over the deodar (cedar) forests of the upper Jamuna valley. These trees had suddenly become market-worthy, to service the then expanding railway network. But as a peasant bitterly observed, "the forests have belonged to us from time immemorial, our ancestors planted them and have protected them; now that they have become of value, government steps in and robs us of them."18 Or consider, finally, these remarks of an administrator in the Bastar district of central India, on the determination of his tribal subjects to continue practicing swidden cultivation in what was now "government" forests:

"On the road from Tetam to Katekalyan I found general dissatisfaction at the restriction of penda [swidden] cultivation. I was unable to convince them of its evils [sic]. Podiyami Bandi Peda of Tumakpal has to get his son married and for this purpose he wants to cultivate penda in the prohibited area. I told him he should not do it. He replied plainly that he would cultivate it and go to jail as he had to get his son married."

In 1871, the Madras Government predicted that the new act, if passed into law, would "place in antagonism to Government every class whose support is desired and essential to the object in view, from the Zamindar [landlord] to the Hill Toda or Korombar." This was an astonishingly accurate prediction, for the Forest Department was unquestionably the most unpopular arm of the British Raj. The story of the numerous popular movements against state forestry, so long neglected by historians, is now attracting an array of chroniclers. The critics were principally of two kinds. On the one side were scholars and politicians with a deep knowledge of rural conditions, and who sometimes formed part of popular movements themselves.

Their criticisms of state forestry thus drew richly upon the feelings and grievances of the people most affected by it. On the other side were the rare (but, for that reason, significant) dissidents within the colonial bureaucracy, who opposed the centralizing thrust of government forest policy. The first set were outsiders so far as the apparatus of rule was concerned, but insiders with respect to popular opinion and popular consciousness. The second set were by virtue of race and status outsiders to Indian society, but insiders with regard to the policy of the state and the functioning of government.

PRECOCIOUS PROPHETS

In 1878, the Poona Sarvajanik Sabha, a vastly respected nationalist organization in western India, bitterly opposed the new Forest Act. Despite its middle-class origins, the Sabha had consistenly fought for the rights of the cultivator, urging that the colonial government lessen its burden of taxation on the peasantry.21 Now, in the context of the Forest Act debate, it pointed out that state usurpation was grossly violative of customary rights over forests, for both "private grantees and village and tribal communities" had "cherished and maintained these rights with the same tenacity with which private property in land is maintained elsewhere." The Sarvajanik Sabha did not, however, merely oppose the proposed Forest Act for its excessive emphasis on state control; it offered a more constructive and creative alternative.

Thus the Sabha argued that the better maintenance of forest cover could more easily be brought about by taking the Indian villager into confidence of the Indian Government. If the villagers were rewarded and commended for conserving their patches of forestlands, or for making plantations on the same, instead of ejecting them from the forestland that they possess, or in which they are interested, emulation might be evoked between neighboring villages. Thus more effective conservation and development of forests in India might be secured, and when the villagers have their own patches of forest to attend to, government forests might not be molested. Thus the interests of the villagers as well as the government can be secured without causing any unnecessary irritation in the minds of the masses of the Indian population.

The Sabha was advocating a far more democratic structure of forest management than that envisaged by the colonial government. Indeed, it was proposing the institution of a Vrikshamitra (Friends of the Trees) Award, one hundred and ten years before the Indian Government's Ministry of Environment and Forests conceived and named such a scheme, for rewarding individuals and communities who had successfully protected or replenished forest areas.

Three years after the 1878 act was passed, the impact of state forestry on rural communities was foregrounded by the social reformer Jotirau Phule. Phule himself was a gardener by caste, and in general exceptionally alert to the problems of the agricultural classes. The following is his description of the impact of the Forest Department on the livelihood of farmers and pastoralists in the Deccan countryside: In the olden days small landholders who could not subsist on cultivation alone used to eat wild fruits like figs and jamun and sell the leaves and flowers of the flame of the forest and the mahua tree [both common trees of the Indian forest].

They could also depend on the village ground to maintain one or two cows and two or four goats, thereby living happily in their own ancestral villages. However, the cunning European employees of our motherly government have used their foreign brains to erect a great superstructure called the forest department. With all the hills and undulating areas as also the fallow lands and grazing grounds brought under the control of the forest department, the livestock of the poor farmers do not even have place to breathe anywhere on the surface of the earth.

These remarks drew attention to the dependence of the agriculturist on the produce of forests and other common lands. This dependence was even more acute in the tribal regions of middle India, where communities of hunter-gatherers, swidden agriculturists, and charcoal iron makers were likewise at the receiving end of the new forest laws. These peoples found an eloquent spokesman in Verrier Elwin (1902-1964), a brilliant Oxford scholar and renegade priest who became the foremost interpreter of adivasi (tribal) culture in India. Elwin was a pioneer of ecological anthropology, whose many works vividly showcased the intimate relationship between the forest world and the life of the adivasi. All tribals, he argued, had a deep knowledge of wild plants and animals; some could even read the great volume of Nature like an "open book." Swidden agriculturists, for whom forest and farm shaded imperceptibly into each other, had an especial bond with the natural world. They liked to think of themselves as children of Dharti Mata, Mother Earth, fed and loved by her.

Elwin's ethnographies are peppered with references to the adivasi's love for the forest. Tragically, the forest and game laws introduced by the British had made them interlopers in their own land. He quotes a member of the tribal group Gond, whose idea of heaven was "miles and miles of forest without any forest guards." As the anthropologist himself wrote in 1941:

The reservation of forests was a very serious blow to the tribesman. He was forbidden to practice his traditional methods of cultivation. He was ordered to remain in one village and not to wander from place to place. When he had cattle he was kept in a state of continual anxiety for fear they should stray over the boundary and render him liable to heavy fines. If he was a Forest Villager he became liable at any moment to be called to work for the Forest Department. If he lived elsewhere he was forced to obtain a license for almost every kind of forest produce. At every turn the Forest Laws cut across his life, limiting, frustrating, destroying his self confidence.

During the year 1933-4 there were 27,000 forest offences registered in the Central Provinces and Berar and probably ten times as many unwhipped of justice. It is obvious that so great a number of offences would not occur unless the forest regulations ran counter to the fundamental needs of the tribesmen. A Forest Officer once said to me: "Our laws are of such a kind that every villager breaks one forest law every day of his life."

Elwin's writings were addressed equally to the colonial state and to the Congress nationalists, who in the 1940s were very much a government-in-waiting. The Congress, however, had not been especially sensitive to the rights of the tribals. But as Elwin reminded them, "the aboriginals are the real swadeshi [indigenous] products of India, in whose presence everything is foreign.

They are the ancient people with moral claims and rights thousands of years old. They were here first: they should come first in our regard." He was deeply distressed when a Congress report on tribals followed the British authorities in asking for a ban on shifting cultivation.

Now Elwin's work had shown that, contrary to modernist prejudice, swidden as practiced by the Baiga, the Juang, and other tribes was an ecologically viable system of cultivation. When the nationalists recommended the ban, he wrote angrily that "the forests belong to the aboriginal. I should have thought that anyone who was a Nationalist would at least advocate swaraj [freedom] for the aboriginal!"

The significance of the forest in tribal life is a running theme in Elwin's work. Noting that a majority of tribal rebellions had centered around land and forests, he pleaded for the greater involvement of tribals in forest management in free India. Even if adivasis had no longer any legal rights of ownership, they had considerable moral rights. And as tribals were as much part of the national treasure as forests themselves, there should be an amicable adjustment between forest management and tribal needs. Even where commercial forest operations became necessary, he said, these should be undertaken by tribal cooperatives and not by powerful private contractors.

After independence Verrier Elwin became the first foreigner to be granted citizenship of free India. In 1954, he was appointed Adviser on Tribal Affairs to the Government of India (with special reference to the North-east Frontier Agency). He was also to serve on more than one high-level, all-India committee on tribal policy. From his first official appointment until his premature death in 1964, Elwin repeatedly urged a reconsideration of forest policy, such that it might, at last, come to more properly serve tribal needs. In this he had little success for forest management became, if anything, more commercially oriented in independent India.32 Towards the end of his life, the anthropologist wrote with some bitterness of how the victims of government policy were being unfairly blamed for the destruction of forests:

There is constant propaganda that the tribal people are destroying the forest. When this was put to some of the villagers, they countered the complaint by asking how they could destroy the forest. They owned no trucks; they hardly had even a bullock-cart; the utmost that they could carry away was a headload of produce for sale to maintain their families and that too against a license. The utmost that they wanted was wood to keep them warm in the winter months, to reconstruct or repair their huts and carry on their little cottage industries. Their fuel-needs for cooking, they said, were not much, for they had not much to cook. Having explained their own position they invariably turned to the amount of [forest] destruction that was taking place all around them. They asked how the exzamindars [landlords], in violation of their agreements and the forest rules and laws, devastated vast tracts of forest land right in front of officials. They also related how the contractors stray outside the contracted coupes, carry loads in trucks in excess of their authorised capacity and otherwise exploit both the forests and the tribal people.

There is a feeling among the tribals that all the arguments in favor of preservation and development of forests are intended to refuse them their demands. They argue that when it is a question of industry, township, development work or projects of rehabilitation, all these plausible arguments are forgotten and vast tracts are placed at the disposal of outsiders who mercilessly destroy the forest wealth with or without necessity.

From a great Englishman who devoted his life to the service of the Indian poor, we move on to a great Englishwoman who did likewise-Madeleine Slade, the daughter of an admiral who came from England to join Gandhi in his Sabarmati Ashram in 1926. Gandhi adopted her as his own daughter and gave her the name Mira Behn.

She played a prominent role in the anti-colonial struggle and was jailed several times.

In 1945, Mira Behn set up a Kisan (peasant) Ashram near the holy town of Hardwar, and two years later moved up the Ganges beyond Rishikesh, where the river descends into the plains. In 1952, she shifted her base again, to the Bhilangna valley in the interior Himalaya. Here she stayed still 1959 when ill health and possibly dissatisfaction with the policies of independent India made her migrate to Austria.

The peasants of the central Himalaya are, of course, as dependent on forest produce as the tribals of the Indian heartland with whom Elwin long worked. Here, one unfortunate consequence of state forest management was the gradual replacement of banj oak (quercus incana), a tree much prized by villagers as a source of fuel, fodder, and leaf manure, by chil (or chir) pine (pinus roxburghii), a species more valued commercially as a source of timber and resin. This transition had serious ecological implications, for the thick undergrowth characteristic of banj forests absorbed a high proportion of the rainwaters of the fierce Himalayan monsoon. This water then slowly percolated downhill. Below the oak forests were thus found "beautiful sweet and cool springs," the main source of drinking water for the hill villagers. By contrast, the floor of pine forests was covered thinly by needles, and had much less absorptive capacity. In hillsides dominated by chil, the rain rushed down the slopes, carrying away soil, debris, and rock, contributing thereby to floods.

Why were the banj forests disappearing in the Himalaya? Mira Behn's own explanation revealed a sharp awareness of the sociology of forest management in the hills. "It is not merely that the Forest Department spreads the Chil pine," she said, "but largely because the Department does not seriously organize and control the lopping of the Banj trees for cattle fodder, and... is glad enough from the financial point of view to see the Banj dying out and the chil pine taking its place. When the Banj trees grow weak and scraggy from overlopping, the chil pine gets a footing in the forest, and once it grows up and starts casting its pine needles on the ground, all other trees die out."

Mira Behn continued: "It is no good putting all the blame on the villagers....

The villagers themselves realize fully the immense importance of these Banj forests, without which their cattle would starve to death, the springs would dry up, and flood waters from the upper mountain slopes would devastate their precious terraced fields in the valleys. Indeed all these misfortunes are already making their appearance on a wide scale. Yet each individual villager cannot resist lopping the Banj trees in the unprotected Government forests. 'If I do not lop the trees someone else will, so why not lop them, and lop them as much as possible before the next comer.'"

Although Mira Behn does not explicitly make the point, it seemed that this shortsighted behavior of the hill peasant was related to the loss of community control, such that individual peasants no longer had a long-term stake in the maintenance of forest cover. This was a tendency aggravated by the commercial orientation of the Forest Department. Could anything be done to restore banj to its rightful place, and thus revive Himalayan economy and ecology? Mira Behn writes: The problem is not without solution, for if trees are lopped methodically, thcy can still give a large quantity of fodder, and yet not become weak and scraggy. At the same time, if the intruding Chil pines are pushed back to their correct altitude (i.e. between 3,000 and 5,000 feet), and the Banj forests are resuscitated, the burden on the present trees will, year by year, decrease, and precious fodder for the cattle will actually become more plentiful.

But all this means winning the trust and co-operation of the villagers, for the Forest Department, by itself, cannot save the situation. Nor can it easily win the villagers' trust, because the relations between the Department and the peasantry are very strained, practically amounting to open warfare in Chil pine areas. Therefore, in order to awaken confidence in the people, some non-official influence is necessary. With the aid of local constructive workers, it should become possible to organize village committees and village guards to function along with the Forest Department field staff which should be increased, and also given special training in a new outlook towards the peasantry.

In this way it should be feasible to carry out a wellbalanced long term project for controlled lopping and gradual return of the Banj forests to their rightful place, by systematic removal of Chil pines above 5,500 feet altitude to be followed by protection of the young Banj growth. The Banj forests are the very centres of nature's economic cycle on the southern slopes of the Himalayas. To destroy them is to cut out the heart and thus bring death to the whole structure.

Mira Behn sent reports of her findings, with photographs, to Prime Minister Jawaharlal Nehru. He passed them on to the concerned officials, but nothing seems to have come of it; it appears that the Indian Forest Department of the 1950s would not change its ways.

A DEMOCRATIZING FORESTER

The quotes offered in the previous section are all strikingly contemporary. To those who know something of the people behind them, they are also perfectly in character. Phule, Elwin, and Mira Behn, as well as the leaders of the Poona Sarvajanik Sabha, had a deep knowledge of agrarian life. Alert to the inequities in access to natural resources brought about by the new laws, they would vigorously polemicize on behalf of the victims of state forest management. Dietrich Brandis was a prophet of community forestry who came from the unlikeliest of backgrounds. He was a forest officer; in fact, no less than the first inspector general of forests (IGF) in India. In nineteen years as IGF (1864-1883), he laid the foundations of state forestry in India.

A man of great energy, he toured widely in the subcontinent, writing authoritative reports on the direction forest management should take in the different provinces of British India.36 In the realm of silviculture, he formulated the systems of valuation and forest working still widely in use. As a former university don himself (he came to the service of the Raj from the University of Bonn), Brandis started a college for training subordinate staff, arranged for higher officials to be trained on the continent, and helped set up the Forest Research Institute in Dehradun.

The scientific and administrative aspects of Brandis's legacy are not our focus here, but rather his sociology of forest management, his understanding of the social and political contexts within which state forestry had to operate in India.37 Here Brandis's views must be immediately distinguished from almost all other forest officials, Indian or European, before or since. These officials counterpose "scientific" forestry under state auspices to the customary use of forests by rural communities, which they have always held to be erratic, unsystematic, wasteful, and shortsighted. It is thus that the forest officials justify their territorial control of over one-fifth of India's land mass, claiming that they alone possess the technical skills and administrative competence to manage woodland.

To be sure, Dietrich Brandis shared this creedal faith in the scientific status of sustained-yield forestry. He also believed that the state had a central role to play in forest management. But what he certainly did not share was his colleagues' skepticism of the knowledge base of rural communities. For example, Brandis wrote appreciatively of the widespread network of sacred groves in the subcontinent. These he termed, on different occasions, the "traditional system of forest preservation" and examples of "indigenous Indian forestry." In his tours he found sacred woodlands "most carefully protected" in many districts-from the Devara Kadus of Coorg in the south to the deodar temple groves in the Himalaya.

At the other end of the social spectrum, Brandis also wrote appreciatively of forest reserves managed by Indian chiefs. He was particularly impressed by the Rajput princes of Rajasthan, whose hunting preserves provided game for the nobility as well as a permanent supply of fodder and small timber for the peasantry. The British stereotype of the Indian Maharaja was of a feckless and dissolute ruler, but as Brandis pointed out, in strenuously preserving brushwood in an arid climate the Rajputs had "set a good example, which the forest officers of the British government would do well to emulate."

In Brandis's larger vision for Indian forestry, a network of state reserves would run parallel to a network of village forests. The Forest Department would take over commercially valuable and strategically important forests, while simultaneously encouraging peasants to collectively manage areas left out of these reserves. Through a series of reports and memoranda written over a decade, the IGF tried to persuade the colonial government that a strong system of village forests was vital to the longterm success of state forestry itself.

The first such report was written in 1868, and pertained to the southern province of Mysore. This was a closely argued document suggesting the creation of village forests throughout Mysore, managed on a rotational cropping system, with freshly cut areas closed to fire and grazing. Ideally, each hamlet would have its own forest, but in many cases it might become necessary to constitute a block to be used by a group of villages. Such forests would provide the following items free of cost-firewood for home consumption and for sale by "poor people with headloads"; wood for agricultural implements and the making and repairing of carts; wood, bamboo, and grass for thatching, flooring, and fencing; leaves and branches for manure; and grazing except in areas closed for reproduction. On the payment of a small fee, wood would be made available for houses and for use by artisans.

In Brandis's scheme, these forests would be put under a parallel administrative system, with a village forester for each unit, a forest ranger for all the village forests in each taluk (county), and a head forest ranger for the district as a whole, this man reporting to an assistant conservator of forests. He anticipated that the system would be self-supporting, with any surplus used for local improvements. In this manner, peasants would come to feel an interest "in the maintenance and improvement of their forests." Brandis also hoped that Forest Department control over village forests would give way in due course, with the "leading men" of each village assuming responsibility for management.

Forwarding his report to the Government of India, Brandis noted significantly that it was "the first of a series of measures" which he proposed "to suggest in various Provinces for the better utilization and for the improvement of the extensive wastelands which will not be included in the State Forests": that is, as a prelude to recommending a countrywide system of community forests. Unhappily, the British officials of the Raj lacked Brandis's understanding of the biomass economy of rural India, the vital dependence of agrarian life on the produce of the forests. They also lacked his faith in local knowledge and local initiative.

The opposition to Brandis claimed that his scheme would lead to a loss of state revenues while undermining the powers of district officials. Also invoked was an early version of the "Tragedy of the Commons" argument. For one official, "the village communities of Mysore, without cohesion and often split up into factions by caste, could not be entrusted with the powers, or competent to perform the functions assigned to them in [Dr Brandis's] scheme." Another commented that the scheme would fail "as each man, when the least removed from supervision, would cut whatever he might require for himself without any regard to the interests of his neighbours." The Government of India's final, negative verdict rested on a classic piece of colonial stereotyping. "The prejudices and rivalries of Natives," it said, "might be excited if men of different classes and castes shared in the same forests."

Brandis did not lightly accept this judgment. In a defiant note, he reviewed the case afresh, and made another forceful plea in favor of village forests. He drew pointed attention to the flourishing system of community forests on the continent, where scientific foresters exercised technical supervision over woodland managed for the exclusive benefit of villages and small towns. In Europe, wrote Brandis, "Such Communal Forests are a source of wealth to many towns and villages in Italy, France and Germany; property of this nature maintains a healthy spirit of independence among agricultural communities; it enables them to build roads, churches, school-houses, and to do much for promoting the welfare of the inhabitants; the advantages of encouraging the growth, and insisting on the good management of landed communal property, are manifold, and would be found as important in many parts of India as they have been found in Europe."

Following his failure in Mysore, Brandis resurrected his proposals in the debate leading up to the 1878 Forest Act. Here he urged the administration "to demarcate as state forests as large and compact areas of valuable forests as can be obtained free of forest rights of persons," while leaving the residual area, smaller in extent but more conveniently located for their supply, under the control of village communities. He hoped ultimately for the creation of three great classes of forest property, based on the European experience: state forests, forests of villages and other communities, and private forests. State ownership had to be restricted, argued Brandis, on account of the "small number of experienced and really useful officers" in the colonial forestry service and out of deference to the wishes of the local population. For "the trouble of effecting the forest rights and privileges on limited well-defined areas is temporary and will soon pass away, whereas the annoyance to the inhabitants by the maintenance of restrictions over the whole area of large forest tracts will be permanent, and will increase with the growth of population."

Here was an uncanny anticipation of the widespread popular opposition that has been such a marked feature of the subsequent history of Indian forestry. But Brandis was overruled by more powerful civil servants within the colonial bureucracy, and the 1878 Forest Act was based firmly on the principle of state monopoly. But the German forester was a remarkably persistent man. As he remarked shortly after relinquishing the post of inspector general, systematic forestry in India "was like a plant of foreign origin, and the aim must be to naturalize it."

On the social side, this process of indigenization could be accomplished by encouraging native chiefs, large proprietors, and especially village communities to develop and protect forests for their own use. In the last instance the initiative lay with the government, which, insisted Brandis, stood to gain enormously from a successful system of communal forests. "Not only will these forests yield a permanent supply of wood and fodder to the people without any material expense to the State," he wrote, "but if well managed, they will contribute much towards the healthy development of municipial institutions and of local self-government."

In 1897, well into his retirement in Germany, Brandis returned to the subject of community forests. Long after he had severed all formal contacts with British India, Brandis continued to be deeply concerned that Indian forestry should cease to have "the character of an exotic plant, or a foreign artificially fostered institution." This concern was consistent with his larger democratic vision for forestry in the subcontinent. Thus he suggested that "native" forest officers, as they distinguished themselves, be sent to study the forestry system operating in Germany. Notably, Brandis had in mind their social as well as silvicultural education. As he concluded his essay, Indian foresters, if sent to Germany, "will find that the villages, which own well-managed communal forests, are prosperous, although now and then they complain of the restrictions that a good system of management unavoidably imposes. What Indian forest officers will learn in this respect in Germany will be really useful to them in India."

Perhaps by now Brandis despaired of British officials in India taking seriously his proposals for the constitution of village forests. Hence this indirect approach, wherein Indian forest officers trained on the continent might be able to better see the benefits of community forests. In the event, Indian officials (whether trained in Germany or not) have been, for the most part, hostile to any suggestion that local communities could be encouraged to manage forest areas for their own use. It is, indeed, this territorial monopoly and indifference to the demands of rural communities that have made the Forest Department the object of such relentless criticism in recent years.

The Indian Forest Department has been the subject of sharp attack for its authoritarian style of functioning; and yet, in an interesting paradox, the founder of the department had himself anticipated that a narrow reliance on state control and punitive methods of management would lead to popular disaffection. While terms such as "social forestry," "community forestry," and "joint forest management" have only now come into currency, the principles they embody would have been readily recognized, and indeed warmly commended, by the first head of the Forest Department in India.

THE HIMALAYAN CASE

For all their insight, knowledge, and passion, these precocious advocates of community forestry did not have much impact on state policy. Control and commercialization remained the dominant motifs of state forest policy. The chapter on village forests in the 1878 act remained a dead letter. Government forest policy, in the colonial as well as postcolonial periods, continued to seriously ignore village needs, demands, and interests. The principle of state monopoly has remained paramount, with one very partial exception.

The Kumaun and Garhwal hills of present-day Uttar Pradesh contain the best stands of softwood in the subcontinent. These coniferous species have been highly prized since the early days of colonial forest management. Between 1869 and 1885, for example, some 6.5 million railway sleepers made from deodar (cedrus deodarus, the Indian cedar) were exported from the valley of the Yamuna, in the princely state of Tehri Garhwal. Adjoining Tehri Garhwal to the east was the British-administered Kumaun division, with its rich stands of chir pine. Here forestry operations concentrated simultaneouly on expanding the area under chir (at the expense of oak) and exploiting the tree both for timber and for resin. Between 1910 and 1920, for example, the number of trees tapped for resin increased from 260,000 to 2,135,000.

The pine trees of the Central Himalaya were the only source, within the British Empire, of oleo-resin, an extract with a wide range of commercial and industrial applications. Likewise, the timber of deodar and chir, as well as fir and spruce, constituted a strategically valuable resource for the colonial state, exploited with profit to service the military campaigns of the two world wars.

In the Himalaya, as elsewhere, commercial forestry under state auspices was made possible only through a denial of customary rights of ownership and use. In these hills, forests and grassland were a crucial resource for the agro-pastoral production system.

In fact, the fragmentary evidence available to historians does suggest the existence of a fairly widespread system of common property resource management-with grass reserves walled in and well looked after, oak forests managed by the village community, and sacred groves lovingly protected. Not surprisingly, the government's attempts to seize vast areas under local control and reconstitute them as "reserved forests" evoked opposition. In the early years of state management, a petition from a discontented hillman evocatively recalled a golden age when the villagers had full control over their forest habitat:

In days gone by every necessities of life were in abundance to villagers than to others [and] there were no such government laws and regulations prohibiting the free use of unsurveyed land and forest by them as they have now. The time itself has now become very hard and it has been made still harder by the imposition of different laws, regulations, and taxes on them and by increasing the land revenue. Now the village life has been shadowed by all the miseries and inconveniences of the present day laws and regulations.

They are not allowed to fell down a tree to get fuels from it for their daily use and they cannot cut leaves of trees beyond certain portion of them for fodder to their animals. But the touring officials still view the present situation with an eye of the past and press them to supply good grass for themselves and their [retinue] without thinking of making any payment for these things to them who after spending their time, money and labour, can hardly procure them for their own use.

In short all the privileges of village life, as they were twenty years ago, are nowhere to be found now, still the officials hanker after the system of yore when there were everything in abundance and within the reach of villagers.When such protests went unheeded, the sentiments underlying them were to manifest themselves in sustained and organized resistance on the part of the Himalayan peasantry. In fact, this region probably witnessed more and more serious social conflict than any other forest region of India.

There were major peasant movements against state forest policies in 1904, 1906, 1916, 1921, 1930, and 1942. These recurrent conflicts, remarked one sensitive official, were a consequence of "the struggle for existence between the villagers and the Forest Department; the former to live, the latter to show a surplus and what the department looks on as efficient forest management."

The most significant forest movement in Kumaun and Garhwal took place in 1921. This took the shape of labor strikes, which crippled the administration, and the widespread burning of pine forests. A total of 395 recorded fires burnt an estimated 246,000 acres of forest. Hundreds of thousands of resin channels were destroyed.

Constituting a direct challenge to the state to relax its control over forest areas, these protests enjoyed enormous popular support, which made it virtually impossible for the administration to detect the people responsible for the fires.

The fires were generally directed at areas where the state was at its most vulnerable, for example, compact blocks of chir forest worked for timber and/or resin. Significantly, there is no evidence that the large areas of broad-leaved forest, also controlled by the state, were at all affected. Thus arson was not random but carefully discriminating-it spared those species more useful to the village economy.

In the vanguard of the 1921 movement were soldiers who had fought for the British in the First World War. Kumaun and Garhwal had long supplied hardy and exceptionally brave soldiers for the British Army-indeed, three of the five Victoria Crosses awarded to Indians between 1914 and 1918 went to this region.

These former men in uniform saw the forest regulations as a bitter betrayal of their interests by the white overlord for whom they had so recently risked their lives. Their protests alarmed the colonial state, for apart from being a reservoir of able-bodied men whom it hoped to continue to recruit for the wars it had to fight, the Kumaun hills bordered both Nepal and Tibet-regions not under direct British suzerainty but in which it had strong trading and political interests.

In the wake of the popular protests, a magisterial critique of government forest policy was published by Govind Ballabh Pant. Pant was a rising lawyer from a peasant household in Almora who went on to become one of the foremost of Indian nationalists, after independence taking office successively as chief minister of Uttar Pradesh and home minister of the Government of India.

His 1922 booklet The Forest Problem in Kumaun described the "burial of the immemorial and indefeasible rights of the people of Kumaun," buried, that is, "between the property-grabbing zeal of the revenue officers and the exhortations of experts of the forest department." As he put it, with legal precision, "the policy of the Forest Department can be summed up in two words, namely, encroachment and exploitation."

Several decades of a single-minded commercial forestry had led to a manifest deterioration of the agrarian economy: "Symptoms of decay are unmistakably visible in many a village: buildings are tottering, houses are deserted, population has dwindled and assessed land has gone out of cultivation since the policy of [forest] reservation was initiated.... Cattle have become weakened and emaciated and dairy produce is growing scarce every day: while in former times one could get any amount of milk and other varieties for the mere asking, now occasions are not rare when one cannot obtain it in the villages, for any price for the simple reason that it is not produced there at all."

Pant's analysis was rooted in a deep knowledge of the local context. He took it upon himself to combat the charge, commonly levied against the hill peasant, "of reckless devastation," a charge "sedulously propagated by prejudiced or ignorant persons." As he wrote, The spacious wooded areas extending over the mountain ranges and hill sides bear testimony to the care bestowed by the successive generations of the Kumaonies.

All of them are not of spontaneous growth and specially the finer varieties bespeak his labour and instinct for the plantation and preservation of the forest. A natural system of conservancy was in vogue, almost every hill top is dedicated to some local deity and the trees on or about the spot are regarded with great respect so that nobody dare touch them.

There is also a general impression among the people that everyone cutting a tree should plant another in its place.... Grass and fodder reserves are maintained, and even nap [cultivable] lands are covered with trees, wherever, though in few cases, such land could be spared from the paramount demand of cultivation. Special care is also taken by the villagers to plant and preserve trees on the edges of their fields.

From this analysis, the solution logically offered was to give back to the peasants the woodland that they traditionally regarded as being within their village boundaries.

"If the village areas are restored to the villagers, the causes of conflict and antagonism between the forest policy and the villagers will disappear, and a harmony and identity of interests will take the place of the distrust, and the villager will begin to protect the forests even if such protection involves some sacrifice or physical discomfort." Pant envisaged that these areas would be under the control of the village panchayats, or councils, under whose direction the "natural system of conservancy" would once again come to the fore. As he shrewdly observed, "some restrictions will be there, but these will proceed from within, and will not be imposed from without."

Clearly, Pant drew upon and systematized the knowledge, perceptions, and analysis of the peasant folk of the Kumaun Himalaya. Thus, after the popular protests in 1921, the Government of the United Provinces set up a Kumaun Forest Grievances Committee. This committee toured the hills, examining some five thousand witnesses in all.

Peasant activists submitted dozens of petitions to the committee, on behalf of individual villages. These identified blocks of forest near every village, where peasants would have exclusive rights of fuel and fodder collection, timber for building, wood for ploughs, bamboos for basket making, etc.

It was being proposed that villagers should have full rights over these forests, which they would manage through their own panchayat.

Based on the evidence it collected, the committee finally concluded that "any attempt to strictly enforce these [forest] rules would lead to riots and bloodshed."

It thus divided the existing reserved forests into two categories-Class I, which were to be managed not by the forest officials but by the civil administration (in theory more sympathetic to rural needs), and Class II, constituting the commercially valuable wooded areas, which were to remain with the Forest Department. It also recommended that the government consider the constitution of village forests as per the demands of the people of Garhwal and Kumaun.

Bureaucracies move at their own pace, and only in 1930 the rules were passed allowing for the formation of van panchayats, or village forests, in the hill districts of the United Provinces. These allowed for a forest patch to be handed over to a village if it lay within its settlement boundaries, and if more than one-third of its residents had applied for permission to the deputy commissioner (DC). Once the DC gave the go-ahead, then the villagers elected, by voice vote, a council (panch) of five to nine members. This council in turn elected a head (sarpanch) among themselves. The van panchayat was empowered to close the forest for grazing, regulate cutting of branches and collection of fuel, and organize the distribution of forest produce.

It could appoint a watchman, whose salary would be paid by villagers' contributions to the panchayat. The panchayat could levy fines, although if the offender did not pay it had then to go to the civil courts for redressal.

The felling of trees, however, required the permission of the Forest Department. The department also claimed 40 percent of the revenue from any commercial exploitation. Of the rest, 20 percent would go to the zilla parishad (district council), with the balance 40 percent kept with the DC on behalf of the van panchayat, which with that official's written consent could use the funds for roads, schools, and other local improvements.

There are now in excess of 4,000 van panchayats in Kumaun and Garhwal, covering an area of just less than half a million hectares. An official report of 1960 remarked that many of these village forest councils had done "exemplary work in connection with forest protection and development." A more recent survey has concluded that the panchayat forests are often in a better condition than the reserved forests. Of twenty-one panchayats surveyed in three districts, the forest stock in thirteen of these were in good condition, in four in medium condition, in three in poor condition.

The researcher concluded that van panchayats have, by and large, maintained oak forests very well, especially in contrast to the dismal condition of the reserves (except for those reserves distant from habitations). The position in respect of chir forests is not so clear, but these seem to have done about as badly under van panchayat control as in the reserves. Various studies suggest that, overall, panchayat forests seem to be in as good or better condition than the reserves.

The van panchayat system constitutes the only network of village forests mandated by law in all of India.61 The concession was made by a colonial state worried of losing control in a sensitive and strategically important border region, and it was not to be replicated elsewhere. After independence, the van panchayat regime was not extended to the adjoining region of Tehri Garhwal, where Mira Behn worked in the 1950s, and where it might have very well contributed to preserving and enhancing the oak forests. Within Kumaun, too, there is considerable resentment over the curbs placed on the autonomous functioning of van panchayats. Though technically under the control of the villagers, the Forest Department can veto schemes for improvement, while of the revenue generated, 40 percent is swallowed by the state exchequer. Forty percent of the rest is by law granted to the village, but this money too first finds its way into a "consolidated fund" controlled by the DC, to which individual panchayats have then to apply.

There are signs of an emerging movement to do away with these constricting rules, to make the management of the panchayats come fully under the control of the villagers. A chronicler of this discontent, himself quite aware of the long history of forest-related protests in Kumaun and Garhwal, writes that "those who know the history of forest struggles say that... the van panchayat movement will be the biggest such movement in the hills."

TWO CHEERS FOR JOINT FOREST MANAGEMENT

From the inception of state forestry in India, perceptive critics have argued for a democratization of resource control, for a correction of the commercial bias promoted by successive governments, and for a proper participation in management and decision making by local user groups. Arguments first offered in the 1870s, and reiterated in subsequent decades, were revived, or reinvented, in the 1970s by the now-famous Chipko movement. It is no accident that Chipko originated in Garhwal and Kumaun, the part of India that has seen some of the most intensive conflicts between the state and the peasantry over forest resources.

The 1970s were marked by a series of forest movements in different parts of India. These took place in the Himalaya, in the Western Ghats, and, above all, in the vast tribal belt extending across the heart of peninsular India.

In the Chotanagpur plateau, forest protests formed an integral part of the larger movement for a separate tribal homeland of Jharkhand, carved out from the huge, unwieldy, and predominantly non-tribal state of Bihar. In one much celebrated case, tribals demolished a plantation of teak, a highly prized furniture wood, that was coming up on land previously under the sal tree (Shorea robusta), a species of far greater benefit to the local economy.

Their slogan, "Sal means Jharkhand, sagwan (teak) means Bihar" was a one-sentence critique of the narrow commercial ends of state forestry. Since the 1970s, there has been an ongoing, nationwide debate on forest policy in India, a debate fuelled by the continuing social tension in forest areas and the evidence of massive deforestation provided by satellite imagery.

This debate has passed through three distinct if chronologically overlapping phases. The first phase might be designated the "politics of blame." The activists speaking on behalf of disadvantaged groups have held the forest officials responsible for environmental degradation and popular discontent. The officials, in turn, have insisted that growing human and cattle populations are the prime reason why fully half of the 23 percent of India legally designated as "forest" was without tree cover.

The forestry debate of the 1970s and the 1980s drew, at times, on the heritage of earlier movements and critiques. The peasants of Garhwal and Kumaun, as this writer found out while doing field work there in 1982-83, were acutely conscious of how Chipko itself drew on a long and honorable history of peasant resistance to state forestry.

Tribal activists in Madhya Pradesh and Bihar, meanwhile, were not unfamiliar with the work and message of Verrier Elwin. And in the villages of the Deccan, social workers liked to offer the same quote of Jotirau Phule's reproduced earlier in the paper, as proof that in the agro-pastoral system of that region, proper access to forests and pasture was vital to survival, and that it was the "great superstructure" of the Forest Department that continued to deny herders and farmers this access. However, perhaps the most direct connection between the past and the present of forest management was effected in the summer of 1982, when the Government of India circulated a new draft forest act. Activists and academics joined hands to demonstrate how the proposed legislation was solidly based upon and, indeed, took further forward the centralizing thrust and punitive orientation of the notorious Indian Forest Act of 1878. After a countrywide campaign, the draft bill was finally dropped by the state.

As tempers cooled and polemic exhausted itself, a second phase, the "politics of negotiation," originated. In villages and state capitals, forest officers and their critics found themselves at the same table, talking and beginning to appreciate, if not fully understand, the other's point of view. Concessions were made by each side, protests suspended by one, and leases of forest produce to industry cancelled by the other.

One product of the growing dialogue between activists and bureaucrats was the approval, by the Indian Parliament in 1988, of a new National Forest Policy. Where the ruling Forest Policy of 1952 had stressed state control and industrial exploitation, the new document instead emphasized the imperatives of ecological stability and peoples' needs.

Then, slowly and hesitatingly, commenced the third phase, "the politics of collaboration." In the state of West Bengal, for example, the Forest Department initiated remedial action on its own, abandoning its traditional custodial approach by inviting peasants to cooperate with it. Thousands of village forest protection committees were constituted, each of which pledged to protect nearby forests in collaboration with the state. Thus previously authoritarian government officials joined with previously suspicious villagers to succesfully regenerate the degraded sal forests of southwestern Bengal.

The success of "Joint Forest Management" or JFM in West Bengal has encouraged scholars, activists, and sympathetic civil servants to demand its replication in other parts of India. Outside its original home, however, the progress of JFM has been slow. Administrative styles and cultures of governance vary widely among the states and regions of India. So do individual orientations, with some forest officials still loathe to relinquish control, while others have been inspired to start village protection committees on their own.

A mapping of the forestry debate in contemporary India would therefore show significant regional variations. Some states are still stuck in the "politics of blame"; others have moved tentatively to the "politics of negotiation." West Bengal and parts of Andhra Pradesh, Madhya Pradesh, and Himachal Pradesh have instituted the "politics of collaboration" through the creation of JFM regimes. In this last scenario there is abundant scope for improvement. As analysts have shown, the JFM model now promoted by the Government of India reflects and sometimes reinforces inequities within rural society.

Gender and caste are two axes of discrimination, with women and low-caste members of the village community not having adequate representation or voice in the decision-making process (this is also true, to a great extent, of the van panchayats in Kumaun.) Likewise, pastoral groups and artisans, who have legitimate claims on forest resources, are sometimes given short shrift. Moreover, the forest officials still claim a monopoly of "scientific expertise," refusing to entertain villagers' own ideas on species choice, spacing, or harvesting techniques.

One serious problem with the JFM model, as currently promoted by the state and donor agencies, is that it allows the constitution of village forest committees only on forestland with less than 40 percent crown cover. This is a deeply constricting rule, which reserves to the state, and the state alone, exclusive rights over the bestclothed lands of India.

Thus forests situated close to hamlets cannot come under JFM regimes if they have more than 40 percent tree cover. Again, the regulations, strictly interpreted, would mean that if local communities were to effectively protect and replenish degraded lands, such that the crown cover was to come to exceed that magic figure of 40 percent, the state could step in and remove the area from JFM-which would be a bizarre outcome indeed.

Nor have changes in policy and orientation been accompanied by concomitant changes in legislation. Thus, the present regime is not flexible enough to allow for spontaneous community-initiated forest regimes to exist along with more orthodox JFM regimes. In some parts of India, the Forest Department is casting a covetous eye on areas well protected by village communities.

Thus in the Uttar Pradesh hills, the old established panchayat forests, managed by villagers, are sought to be brought under the JFM system only so that bureaucrats would have a greater say in their management. A new, carefully thought out Indian Forest Act is called for, which allows both for areas to be managed under state-village partnerships as well as by self-generated, autonomous community regimes.

One can thus envision a fourth (and possibly final) phase for the Indian forestry debate, the "politics of partnership." For collaboration, even where it does exist, takes place on terms set down by the state, through the officials of the

Forest Department. We need to move on to a more inclusively democratic structure, where the state listens to and learns from the community, and where the community itself recognizes and deals fairly with the inequities within its own ranks.

The evidence suggests that contemporary advocates of decentralized forestry have had far greater success than their precursors. One reason for this is the altered political context: Brandis, Pant, and company worked under a colonial, authoritarian regime; the partisans of Chipko and similar movements in a democratic system.

The revival of forest protest in the 1970s also coincided with the international environmental debate, which foregrounded the use and abuse of forests worldwide. The work of Indian scholars had, meanwhile, demonstrated with authority that the century-old history of state forestry in India must be reckoned a failure, in both an ecological and social sense. Finally, the problems with government-directed development programs in much of the Third World had led to an increasing interest in nongovernmental forms of management and control.

These calls for forest reform from the outside were complemented by pressures from change from within. Starting with West Bengal, the governors themselves, namely the forest officials in charge of their vast landed estate, realized that old methods of control and exclusion were merely fuelling social conflict. An overworked and underfunded bureaucracy then started, slowly, to involve communities in forest working. What started as a strategic imperative became, at least for some forest officers, a sincere change of heart.

Once the critics from without were being echoed by the dissidents from within, the process of reform accelerated. This is indeed the signal lesson of Indian forest history-that meaningful policy change comes about only when the sustained pressure by social movements and their intellectual sympathizers resonates with the feelings of powerful officials within the state bureaucracy.

One or the other, by itself, will not do. When Brandis was active, he was handicapped both by his lone dissident voice within the Forest Department, and by the fact that there had not yet emerged an effective critique from outside. When Elwin, Mira Behn, and others propagated the feelings and aspirations of the peasants and tribals they worked with, the forest bureaucracy was, collectively and to the last man, deaf to their arguments and entreaties.

Forest policy remained unbending and unchanging, with the exception only of the Kumaun hills. There, as we have seen, the popular protests and outside critics were partially successful not because of a honest rethinking by the state but by its concern that this sensitive border region must not be tempted into outright rebellion. Elsewhere, where this political imperative did not come into play, the colonial regime refused to heed the widespread criticisms of its system of forest exploitation.

In more recent times, however, the radical critics have been aided by the autocritique of influential sections of the forest establishment. This confluence of external pressure and internal rethinking explains why, and how, the contemporary proponents of community forestry have, unlike their predecessors, been able to see their ideas and polemic become translated into official policy and (though less assuredly) into official practice. Nonetheless, there are indeed striking parallels between the ideas underlying the application of joint forest management today and the ideas of the early, prescient, and brave but for the most part unheard critics of state forestry discussed in this essay. With respect to the role of forest dependent communities, for example, there is a shared faith in indigenous knowledge, in the management capacity and robustness of local institutions, and above all, a sharp focus on local access to the usufruct of the forests.

Again, with respect to the role of the state, there is a common recognition of the essentially advisory role of the forest department, of its need to collaborate with rather than strictly regulate customary use, and of the justice of sharing revenues from forest working with the villagers.

Then, and now, critics have called strongly for an attitudinal change among state officials, a retraining and retooling in keeping with the democratic spirit of the age. Finally, both past and present proponents of decentralization seem to converge in their larger vision for forest policy in India, a vision which in my understanding consists of three central principles: (1) that benefit sharing (between state and community) and local control are to be the key incentives to ensure sustainable management and minimize conflict; (2) that community-controlled forests would work as a complement to a network of more strictly protected areas, further from habitations, that continue under more direct state control; (3) and finally, that the restricting of state control to these latter areas is vital on grounds of equity (i.e., the respect for local rights and demands), efficiency (i.e., as the most feasible course, with the state not biting off more than it can chew), and stability (i.e., as the most likely way to lessen conflict).

There is little question that the ongoing attempts at reversing or mitigating state monopoly over forest ownership and management do constitute a significant departure from past trends. In a deeper sense, however, contemporary attempts at fostering participatory systems of forest management hark back to a much older tradition. In the late twentieth century, as in the late nineteenth century, there has arisen a movement for the democratization of forest management, for a system founded not on mutual antagonism but on genuine partnership between state and citizen. The first inspector general's vision for Indian forestry was abruptly cast aside in the 1860s and 1870s, but it may yet come to prevail. That would be a vindication of the life and work of Dietrich Brandis, but also of Jotirau Phule, Verrier Elwin, Mira Behn, Govind Ballabh Pant, and the Poona Sarvajanik Sabha. Ramachandra Guha is an author and columnist based in Bangalore. His books include The Unquiet Woods (first published 1989; second revised edition, University of California Press, 2000), and Environmentalism: A Global History (Addison Wesley Longman, 2000). He is now working on a history of independent India.

FUELWOOD AND ENERGY

In spite of the relatively high per capita income levels in Antigua and Barbuda, there is still some consumption of charcoal. Unfortunately, no reliable records exist to indicate the levels of production and consumption. Charcoal burning does not presently constitute a threat to forest resources as suitable material is already available from pastures in the form of Acacia. This tree is a pest which threatens the health of cattle, as the thorns cause severe skin lesions, exposing the animals to the risk of disease. The GOAB is obviously aware of the continuing world energy crisis, and a case could therefore be made for encouraging the use of alternate sources of energy to save foreign exchange and to utilize natural products generated in the course of development initiatives such as agroforestry and watershed rehabilitation.

Conservation of Forest Ecosystems

In spite of its small size, the island possesses considerable wildlife and ecosystem diversity, much of which could enhance the growth of tourism but, more importantly, could awaken society's consciousness and pride in its natural environment. The Country Environmental Profile has recommended specific elements to be included in a system of national parks and protected areas, and these should be expanded to include coastal zones and forested areas.

Negative development pressure on the coastal zone can compromise the future ability of the country to sustain long-term socio-economic growth. All development proposals should be carefully reassessed to avoid adverse environmental impact. In the meantime, interim action can be taken to stimulate financing of resource management through collection of fees by local guides and monitoring of visitor levels and impact. Resource inventories are proposed under the NFAP projects.

Institutions

The main institutions concerned with forestry and environmental management include the Forestry and Fisheries Divisions, the Development Control Authority (DCA), the National Parks Authority (NPA) and NGOs, primarily the Environmental Awareness Group (EAG). Under the present tight fiscal policies, the public agencies are restricted in their ability to execute programmes. There are no funds available to undertake work programmes in forestry or agroforestry, to maintain monitoring of fisheries and fishermen, nor to expand the capability of the NPA.

The NGOs led by the EAG, are particularly knowledgeable and committed, and though lacking in resources, they are quite resourceful. They have played a significant role in raising environmental consciousness, although their confrontational approach has been met with official suspicion. However, they are now widely accepted in official circles that recognize the desirability of closer collaboration. This has led to the appointment of the Historical, Conservation and

Environmental Commission (HCEC), aimed at improved dialogue and coordination within the public sector. However, the general institutional situation of the public sector is weak in implementing capacity and performance, and the NGOs are still skeptical that the Government will change its ways.

The NPA is backed by strong legislation which provides for inter-sectoral linkages to consult and cooperate with agencies having similar aims and objectives. It also has the power to delegate its functions and this could be done through the Barbuda Council, Forestry and Fisheries Divisions, the EAG and other NGOs.

The Barbuda Council was established by the Barbuda Local Government Act (No. 15, 1976), to manage Barbuda's government affairs. The Council consists of 9 members who elect a chairman and secretary from among them. Proposals for capital and special expenditure are required to be submitted to the Ministry of Home Affairs (Central Government) who forward them to the Ministry of Finance. The Council has powers to make bye-laws for development on the Island. The conservation and development of forest resources can only be implemented by capable resource management institutions, properly staffed and financed, and provided with the necessary authority for efficient execution of their responsibilities. Unfortunately, the lead agencies in Antigua and Barbuda do not seem to be adequately equipped at present, and will require considerable strengthening to improve their management capacity.

With the recent return of the country's first professional forester, the Government now possesses the nucleus of trained forestry personnel. The Forestry Division can now be empowered with the following responsibilities and functions:

(a) protection of the forest estate, wildlife sanctuaries and protected areas;

(b) afforestation and forest management in watershed areas;

(c) cooperation with private and Government agencies on urban forestry programmes (including the Botanical Gardens), forestry education and information;

(d) assessment and conservation of biodiversity and promotion of ecotourism;

(e) generation of self sustaining activities utilizing forest and non-wood products;

(f) preparation of budget plans and annual performance reports.

The following steps are therefore recommended for official approval and action:

(i) adoption of a formal national forest policy and appointment of the Chief Forestry and Wildlife Officer;

(ii) declaration and gazetting of the forest estate, followed by survey and demarcation of forest reserve boundaries;

(iii) preparation and approval of a baseline work plan for the Division, including job descriptions for all staff;

(iv) transfer of game wardens, caretakers and necessary staff to the budget of the Division;

(v) assignment of responsibility for supervising the maintenance of the Victoria Park and Botanical Gardens;

(vi) assignment of office accommodation, vehicles and equipment to enable efficient performance of the Divisions functions.

After establishment of the Division, there are a number of basic activities which it can undertake, namely:

- planning, layout and implementation of a forest recreation centre at Wallings;

- planning and implementation of urban forestry based on the Botanical Gardens;
- field assessment of planted forested trees, enumeration, diameter and height measurements, site conditions;
- assessment of charcoal production from leucaena at Dunbars;
- continued cooperation with Fisheries Division and NGOs in mangrove and wildlife monitoring;
- initiation of contacts with NPA to discuss areas of collaboration;
- selection of possible reforestation sites and planning proposals for employment absorption.

These activities can be undertaken with little or no new financing and are within the technical capability of the trained staff. The results will soon be evident if tasks are approached in a business-like manner and with an eye to the future.

PLAN OF ACTION

Based on the results of the Country Mission, as outlined in the previous sections of this report, and on the working documents prepared by various consultants, a Plan of Action is proposed for the development of the country's forestry subsector. The National Forestry Action Plan (NFAP) proposes elements of national development policies, a long-term policy for the sub-sector, together with priority area policies and objectives on which specific action projects are proposed for a five-year period. Mindful of the country's present financial situation, the Plan proposes redeployment of the existing personnel and budget as the Government's initial investment in the sub-sector to attract donor support in setting a sound foundation for sustainability. The plan includes the normal work of the Forestry Division which will also function as lead agency in the implementation of three projects. The NFAP is estimated to cost a total of US $3.0 million of which US $2.2 million would be contributed by international donors while the remaining US $0.8 million would be contributed by Antigua and Barbuda, mainly in kind.

National Development Policies

The adoption of national development policies for land use planning, land allocation and sustainable resources management are recommended for enhancing the contribution which forestry and trees can make to socio-economic development and environmental stability.

Long-term Subsector Policy and Objectives

59. It is proposed that the long-term Forest Policy of Antigua and Barbuda should be:

(a) to protect the country's natural forest resources;

(b) to promote the development of forests and tree cover on suitable lands;

(c) to encourage the application of agroforestry in planned land use for rural development;

(d) to pursue opportunities for utilization of forest and other products.

In pursuing this forest policy, the Government will seek to ensure:

- the wise use and proper management of the island's limited forest and natural resources;
- the preservation of representative samples of the island's natural forest and vegetation resources and the maintenance of the ecological balance;
- the provision of opportunities for the public for scientific, educational and recreational activities to

enhance knowledge of forest ecosystems, research in potential utilization of the gene pool and tourism benefits;

- the stimulation of industries capable of boosting the island's economy, based on products derived from forests and trees.

Therefore, the proposed objective of the NFAP is to establish and manage a national forest estate on Government and private lands. This will ensure conservation of soil and water resources, protection of forest ecosystems and production of wood and related products for social and economic benefits on a sustainable basis.

Priority Area Policies and Projects

In the medium term, the NFAP will be guided by the priority area policies and strategies outlined below in the implementation of the respective action projects. These policies are all inter-related and represent the minimum actions required to effect the fundamental changes necessary towards achieving the basic aims of the new long-term forest policy. Institutions-to provide effective environmental legislation and capable institutions for proper administration and sustained development of forestry and related programmes, with the participation of local people and interest groups. The purpose of this policy is to improve the environmental management-capacity of the country and to install multi-sectoral and participatory mechanisms for planning and monitoring of forestry and related programs.

UTILIZATION OF FOREST PRODUCTS

Timber

Local people use four hardwood timber species (Dipteriocarpus sp., Parashorea chinensis, Parashorea stellata, and Heritiera cochinchinensis) to meet their demands for timber wood (Grever, 2001). The highest demand was observed in those villages where wooden houses are dominant and hardwood is mainly used for house construction and repairing. Wooden houses are usually refurbished once every five to ten years for which a volume of 0.5 m 3 to 1.5 m 3 is required. A lesser amount of timber wood is also used for construction of furniture. Despite the fact that timber wood exploitation from the National Park is strictly prohibited, there are still many cases of forestry law violation. The number of forest violation cases in the two districts of Nam Dong and Phu Loc belonging to the buffer zone, and Bach Ma National Park appears very high, recorded at 4,545 cases between the years of 1995 and 2000. The quantity of wood to be seized recorded at 4,587m

Firewood

One of the most important non-timber product for daily life in the buffer zone is firewood. Each interviewed household requires wood for preparing food because other means of energy use (e.g. gas, charcoal) are not yet widely introduced. The estimated daily firewood demand lies between 6.1 and 21.3 kg and reflects the fact whether local households are involved in traditional pig production. In this system, the typical diet fed to fattening pigs is mainly based on cooked rice and cooked cassava leaves and roots, which increase the daily demands for firewood considerably.

Firewood is usually collected in the forest or the home garden around the house and a share of the firewood is often sold in the local market. Better-off households tend to purchase their supply whereas poorer households are engaged in the collection of firewood.

According to Eve (1997), 20 kg of firewood has a volume of approximately 0.08 m 3 which leads to an estimated annual firewood demand of 8.9 m 3 to 31.03 m 3 per household. Extrapolating these figures to the 12,285 households living in the buffer zone of Bach Ma National Park reveals the annual consumption of 109,410 m 3 to 381,204 m 3 firewood for the whole buffer zone area. This large quantity is not exclusively extracted from the protected forest resource but also from forest plantations, home gardens, and bush fallow in the buffer zone. However, a number of the villages around the National Park do not have access to any compensation areas in order to meet their daily demands for firewood and are prompted to access the protected area regularly.

NON-WOOD FOREST PRODUCTS

Due to their economic value, good accessibility and high availability, rattan and several products deriving from palm trees are the most commonly exploited non-wood forest products in the National Park. Three different species of rattan are commonly harvested: Calamus scipionum, Calamus rudentum, and Calamus tonkinensis; the two latter species are listed as threatened in Vietnam (Phan, 1996). Local people reported the collection of rattan on 15 to 30 occasions a year, mainly in the dry season from June to August when activities in the agricultural sector are reduced (Grever, 2001). Each harvest lasts up to two days and yields a quantity of approximately 50 trunks. The mode of exploitation is destructive because the hole plant is cut on soil level and only the roots are left. Rattan trunks are usually sold to local traders, but occasionally the leaves are also used for self-consumption (e.g. roof cover) as a salvage product (Johnson 1998).

The price for one trunk depends upon both the diameter and the quality of the product, but it is generally in the vicinity of 1,000 VND per trunk. Palm leaves, probably belonging to the genus Livistonia or Licuala, are the raw material for manufacturing the traditional Vietnamese conical hats. The collection of these leaves is strongly associated with access into the forest for the exploitation of rattan. The harvest rate is estimated at 1,000 to 1,500 leaves per harvest which generates an income in the range of 40,000 to 60,000 VND.

Leaf harvesting of the palm Caryota sp. is exclusively used for two different purposes of self-consumption: (i) the leaflets with a length of about one meter are traditionally used for covering the roof of houses and sheds; (ii) stalks are commonly used as a salvage product for building fences and walls along the home garden and agricultural fields. Wild crafting of orchids as well as hunting and collecting of mammals, birds, and frogs were mainly recorded through additional irregular observations and therefore, these data lack of accuracy. The flora of Bach Ma National Park includes at least 1,400 species of which 432 are potentially used by the local population for a variety of medicinal purposes (Tran & Ziegler 2001).

However, this traditional knowledge is mainly found among the ethnic minorities of Ka Tu and Van Kieu, living in remote areas of the buffer zone which was outside the geographical focus of this study. The surveyed two communes were established within the last 25 years and the majority of the interviewees lack the knowledge of how to use the medicinal plant resource systematically.

The current survey indicates that no medicinal plant is collected inside the National Park; origins of the crafted plant resources derive from home gardens, rice paddies and other locations in close proximity to the surveyed village.

Several reported species are multiple-use plants, belonging to commonly growing wild herbs, cultivated fruit trees (e.g. Psidium guajava, Chrysophyllum coinito), vegetables (e.g. Eclipta prostrata, Houttuynia cordata, Centella asiatica), or ornamental plants (e.g. Polyscias fruticosa,Cordyline fructicosa).

FOREST SERVICES

The interviewees were asked to specify marketable and non-marketable services they gather and obtain from the forestry resource in the National Park. Several answers per interviewee were possible. However, more than one third of the interviewees regarded the forestry resource as having no use value and did not associate any environmental service with it. For this group the depletion and the eventual loss of the protected forest has reportedly no immediate impact upon their livelihoods.

DISCUSSION

Establishment of protected areas generally follows restrictions on traditional resource uses by local communities (McNeely, 1988; Gomez-Pompa & Kaus, 1992). Economic losses due to reduction in land and livestock holdings are reported frequently in the vicinity of protected areas. Farmers often cannot compensate these changes due to a lack of technical and financial support from the area's management board or governmental departments (Maikhuri et al., 2001). Income and supplies from forest related activities continue to play a vital role for emergencies during periods of agricultural shortfalls or unemployment (Falconer, 1994; Ogle, 1996) and the subsistence use of firewood and NWFP remains large and essential.

Vietnam also has one of the highest human population densities in Asia (PRB, 1996), which is the source of the primary threats to the country's natural resources and long-term persistence of its biological diversity. The forests of Bach Ma were also subject to a rapid decline in forest cover caused by anthropogenic activities and intensive pressure upon the natural resource base following the resettlement programme in the aftermath of the American War and the introduction of the market economy in 1986. Wilkie & Godoy

(1996) argue that, with increased exposure to trade and markets, use of the forest is increasingly concentrated on. some higher value Non-Timber Forest Products. This development was observed in Bach Ma in the late 1980s when, for example, excessive and destructive harvesting of the valuable resin plant Aquilaria crassna (Thymelyaceae) has diminished the resource and the species is now believed to be locally extinct (Larsen, 1997).

The destructive harvest of rattan has to be carefully monitored, as there is a large rattan processing capacity available within the vicinity of the park in Hue and Da Nang City. The depletion or removal of certain species can rapidly influence such forest characteristics as composition and structure of vegetation (Redford, 1992; Fitzgibbon et al., 1995). These impacts have nowadays created empty forests in Vietnam lacking the intact floral and faunal communities necessary to maintain viable ecological processes (UNDP, 1997).

Generally, people in the buffer zone of Bach Ma National Park appear to be positive towards conservation. A possible explanation for the focus on non-consumptive values may be concluded from the fact that local people in the buffer zone are mainly engaged in subsistence farming where water and soil are both crucial and precious resources whose abundance and quality determine the survival of the farming economy. Furthermore, the severe floods in November 1999 caused several land slides in the area that killed a total of 34 people and destroyed hundreds of hectares of agricultural land in the buffer zone (Phu Loc Statistical Office 2000). It is very likely that this natural disaster raised the level of awareness for the necessity of conserving the forest.

Direct environmental benefits have favored an environment in which only a minor part of the local population still perceives mostly negative aspects of the existing conservation policy. This study has demonstrated that local people attribute with the National Park a resource they value and seek to conserve. Traditionally, local requirements are perceived as a threat to conservation efforts, with the consequence that little thought is given to how local inhabitants could also benefit from the establishment of the protected area (Ghimire, 1994). More emphases on the demands and aspirations of local people could favor a shift from a predominantly protective orientation in forest management and conservation policy towards encouraging sustainable systems for production of livelihoods benefits (Freese, 1997).

BIBLIOGRAPHY

Bahati, J.: (1995). *Impact of Arboricidal Treatments on the Natural Regeneration of Species Composition in Budongo Forest,* Uganda. M. Sc., Makerere University.

Barrows, M.: (1996). *A Survey of the Intestinal Parasites of the Primates in Budongo Forest,* Uganda. Glasgow University.

Bennet, H.: (1996). *Study of Kinship, Social Relations and Possible Avoidance of Incest among Budongo Chimpanzees.* M. A., Edinburgh University.

Fairgrieve, C.: (1995). *The Comparative Ecology of Blue Monkeys (Cercopithecus mitis stuhlmanni) in Logged and Unlogged Forest, Budongo Forest Reserve,* Uganda, Edinburgh University.

Hills, C.: (1998). *An Analysis of Phylogeny and Gene Flow in a Population of Pan troglodytes schweinfuthii.* M. Sc., Oxford University.

Johnson, K.: (1993). *Local Uses of Budongo's Forest Products.* M. Sc., Oxford University.

Johnson, K.: (1996). *Local attitudes towards the Budongo Forest,* Western Uganda. Indigenous Knowledge & Development Monitor.

Miller, P. L.: (1993). *Some dragonflies of the Budongo Forest,* Western Uganda. Opusc.

Owiunji, I.: (1996). *The Long Term Effects of Forest Management on the Bird Community of Budongo Forest Reserve,* Uganda. M. Sc., Makerere University.

Reynolds, V.: (1965). *Budongo: A Forest and its Chimpanzees,* London, Methuen.

INDEX

A

Agriculture, 15, 23, 24, 25, 27, 30, 68, 73, 74, 75, 81, 82, 84, 85, 86, 99, 101, 103, 104, 107, 138, 139, 140, 141.
Agroforestry, 78, 85, 86, 88, 89, 91, 92, 96, 97, 98, 105, 106, 107, 213.
Authority, 24, 25, 37, 50, 55, 58, 60, 70, 72, 92, 93, 102, 106, 110, 197.

B

Business, 18, 95, 213.

C

Communication, 166.
Conservation, 36, 38, 40, 41, 46, 47, 48, 50, 51, 52, 53, 54, 56, 57, 60, 68, 72, 73, 74, 75, 79, 80, 82, 83, 85, 86, 87, 89, 92, 93, 94, 96, 97, 98, 99, 105, 107, 109, 115, 122, 123, 130, 131, 133, 134, 135, 136, 138, 141, 142, 148.
Considerations, 40, 58, 251.
Contribution, 45, 54, 63, 64, 65, 79, 96, 144, 160, 162, 208, 216, 230, 236, 240, 245, 250, 251.
Corporation, 2, 4, 85.
Cross Sectoral Commitments, 11.

D

Department, 7, 8, 25, 27, 29, 36, 113, 114, 118, 119, 135, 139, 145, 146, 148, 169, 170, 171, 172, 173, 174, 175, 176, 179, 180, 182, 183, 186, 188, 189, 191, 192, 193, 195, 197, 198, 199.
Development, 20, 29, 31, 46, 47, 48, 56, 66, 74, 78, 79, 80, 83, 84, 85, 86, 88, 91, 92, 93, 95, 96, 97, 98, 100, 103, 105, 107, 109, 110, 115, 116, 119, 121, 122, 124, 126, 129, 130, 131, 132, 133, 134, 135, 136, 137, 138, 139, 140, 142,

E

Economy, 79, 82, 96, 162, 180, 183, 189, 194, 211, 240, 245, 246, 247, 249, 251.
Ecosystems, 80, 81, 92, 96, 97, 98, 99, 109, 125, 126, 131, 139, 150.
Energy, 11, 12, 91, 92, 98, 107, 181, 216, 251.

Evaluation, 7, 16, 30, 32, 33, 34, 37, 39, 40, 69, 120, 123, 126, 149.

F

Forest Biodiversity, 117, 118, 126.
Forest Conservation, 19, 79, 215.
Forest Legacy Areas, 26, 32, 36, 38.
Forest Management, 6, 10, 13, 17, 20, 24, 34, 52, 59, 90, 94, 114, 115, 116, 117, 119, 120, 123, 124, 125, 126, 127, 128, 129, 133, 140, 145, 146, 148, 149, 150, 167, 168, 174, 177, 178, 179, 181, 186, 188.
Forest Policy, 1, 2, 3, 8, 11, 14, 15, 16, 17, 18, 19, 86, 94, 96, 97, 115, 116, 124, 129, 151, 173, 177, 186, 189, 190, 194, 195, 198, 199.
Forest Project, 112, 212.
Forest Valuation, 20, 163.
Forestry Projects, 4, 9, 14, 17, 162, 202.
Forestry Sector, 81, 116, 247.
Foundation, 70, 71, 88, 95.

I

Implementation, 1, 3, 5, 7, 9, 10, 14, 15, 16, 17, 23, 25, 26, 29, 31, 35, 38, 45, 87, 94, 95, 97, 99, 100, 101, 102, 103, 106, 116, 119, 120.
Improvements, 13, 49, 66, 76, 131, 143, 183, 192, 211, 247, 249.
Information, 15, 16, 21, 33, 34, 35, 39, 41, 48, 49, 52, 65, 66, 68, 75, 76, 80, 94, 103, 119, 134, 135, 136, 146, 159.
Institutions, 63, 92, 93, 97, 102, 124, 143, 144, 147, 150, 185, 199, 223, 236, 243.

M

Methodology, 123, 159, 160, 161, 211, 214, 256.

N

National Forestry, 78, 79, 95/

O

Observations, 104, 107, 110.
Operations, 3, 7, 11, 13, 14, 32, 116, 124, 151, 177, 187, 211, 221, 238, 240, 241, 246.

P

Performance, 10, 14, 15, 55, 90, 93, 94, 107, 214, 245, 249.
Plantations, 3, 5, 9, 11, 19, 81, 122, 174, 211, 239, 241,

R

Research, 31, 83, 96, 99, 103, 120, 125, 138, 164, 200, 251, 257.

S

Significance, 9, 75, 159, 177, 219, 229, 245, 254.

T

Technology, 106, 201.
Treatments, 67, 89, 205.

❑❑❑